# A GUIDE TO PROTEIN ISOLATION,
2nd edition

## FOCUS ON STRUCTURAL BIOLOGY

Volume 3

*Series Editor*

ROB KAPTEIN

*Bijvoet Center for Biomolecular Research,*
*Utrecht University, The Netherlands*

# A GUIDE TO PROTEIN ISOLATION

## 2nd edition

by

CLIVE DENNISON
*School of Molecular and Cellular Biosciences,*
*University of Natal,*
*Pietermaritzburg, South Africa*

KLUWER ACADEMIC PUBLISHERS
DORDRECHT / BOSTON / LONDON

A C.I.P. Catalogue record for this book is available from the Library of Congress.

ISBN 978-90-481-6266-6

---

Published by Kluwer Academic Publishers,
P.O. Box 17, 3300 AA Dordrecht, The Netherlands.

Sold and distributed in North, Central and South America
by Kluwer Academic Publishers,
101 Philip Drive, Norwell, MA 02061, U.S.A.

In all other countries, sold and distributed
by Kluwer Academic Publishers,
P.O. Box 322, 3300 AH Dordrecht, The Netherlands.

*Printed on acid-free paper*

# Contents

## CHAPTER 6



## CHAPTER 7

# Acknowledgements

I am grateful to the reviewers of the 1st edition, and the reviewers of the draft of this edition, for pointing out shortcomings and areas where the book could be improved. In answer to their critiques, I have added more study questions and have provided answers. I have also added a section on HPLC and a chapter on practical methods, in the hope of making the book more useful in the lab. I wish to acknowledge the contribution of Ron Berry, Senior Customer Assurance Specialist, Waters Australia, Melborne. The section on HPLC was distilled, with permission, from a teaching manual which he prepared when he was a staff member in the same Department as myself.

My fundamental philosophy is that whenever one is doing something practical, *you should understand what you are doing!* In the hope of increasing students' understanding, I have added a prefatory chapter, dealing with the physics on which protein isolation methods are based. In writing this chapter, I was obliged to clarify my own thinking on some matters and I wish to thank my colleagues in Physics for answering my sometimes naive questions, for correcting the formatting of equations etc. and for pointing out some conventions in physics, of which I was not aware.

# Preface

It is a truism of science that the more fundamental the subject, the more universally applicable it is. Nevertheless, it is important to strike a level of "fundamentalness" appropriate to the task in hand. For example, an in-depth study of the mechanics of motor cars would tell one nothing about the dynamics of traffic. Traffic exists on a different "level" - it is dependent upon the existence of motor vehicles but the physics and mathematics of traffic can be adequately addressed by considering motor vehicles as mobile "blobs", with no consideration of how they become mobile. To start a discourse on traffic with a consideration of the mechanics of motor vehicles would thus be inapproppropriate.

In writing this volume, I have wrestled with the question of the appropriate level at which to address the physics underlying many of the techniques used in protein isolation. I have tried to strike a level as would be used by a mechanic (with perhaps a slight leaning towards an engineer) - i.e. a practical level, offering appropriate insight but with minimal mathematics. Some people involved in biochemical research have a minimal grounding in chemistry and physics and so I have tried to keep it as simple as possible.

Besides trying to find the right level, I have tried to show that the physical principles which can be employed in protein isolation are, in fact, ubiquitously applicable principles with which students may be well familiar, though perhaps in different contexts. These "ubiquitously applicable principles" - once identified as such - turn out to be old and familiar friends, with whom one can have a great deal of fun when applied to the challenges of protein isolation.

In an uncertain world one never knows what the future will bring - who knows whether the economy, the state of world politics, or the weather, will be better or worse this time next year than it is now? - but one of the enduring attractions of science is that, because of the labours of scientists throughout the world, it is almost certain that, "this time next year we'll have greater understanding and insight". This book is offered in the spirit of sharing some of the insights that I have gained in my career in Biochemistry. In some instances, I might have got hold of the wrong end of the stick. Where this is the case, I would welcome comment so that we might all learn - as we always do - from the errors.

*Clive Dennison*

# Preface to the 2nd edition

In the 1st edition of this book, I made the assumption that the reader would have some background in physics and would at least understand the elementary concepts of Newtonian physics. This edition is aimed at undergraduate students, and my experience has been that students may indeed have attended and passed courses in physics and may indeed have acquired some grasp of the concepts in the context in which they were given. However, many students have difficulty in transferring their knowledge to a new situation. They have a tendency to package their knowledge in mental boxes - one being labelled "physics" - but seldom reopen those boxes to use the material again in a subject which is not called "physics".

One of the attractions of physics is that it greatly simplifies the natural world. Only a few physical principles are necessary to explain almost everything - at least at the scale at which we live our daily lives and at which biochemists operate. Many students try to "learn" the material presented in a course - filling their minds with a large number of unconnected facts which soon become overwhelming and the material is "lost" as soon as the examinations are over and, more importantly, no tools are acquired with which to tackle new, presently-unimaginable problems. In fact, the tools are much more important than the "facts". Who can say what problems current students will face in, say, 20 years time? All we can say with certainty is that technology will change - probably becoming more complex - but the principles of physics will apply and if the erstwhile student has a good grasp of these, they will be able to deal creatively with any new challenges.

One of the challenges facing teachers of science, in my opinion, is to show how the real world is not divided into "boxes". Biochemistry may be taught separately from physics but, in the real world, there are no divisions. Similarly, in the real world, there is no division between the biochemical separations one may do in the lab and one's everyday life - the same physical principles apply. In this book, a recurrent theme, therefore, is to make connections between seemingly-unrelated phenomena, in order to show the connections between every-day events and the things one may do in the lab., i.e. *how the same principles apply!* And the more one sees the bigger picture, the simpler everything becomes.

As most of the separation methods used in protein isolation have a basis in physics, it is necessary to have a good understanding of the physical principles, so that one can properly understand how the separation methods work. For this reason, I have added a prefatory Chapter on some of the relevant principles of physics - as a revision or limbering-up exercise before these principles are applied in the protein isolation methods to follow.

**A note to students.**

The purpose of learning should be to increase one's insight and understanding. However, there is much debate about what is meant by the word "understanding". In some respects it has to do with familiarity. This is because of the way our brains are structured. Because a particular neural network is reinforced by repetitive usage, we become better and better at the things that we do often - "practice makes perfect", as the adage goes. However, we can become very good at doing something, without really understanding what we are doing. We can learn to catch a ball or ride a bicycle, for instance, without knowing Newton's laws of motion. Doing these things requires the development of very quick, but subconscious skills.

Learning by repetition can be corrupted into rote learning, where the object is to retain some knowledge simply by repetition, but without proper, conscious, understanding and without integrating it with one's existing knowledge. To a limited extent this can work - I can still recite parts of some foreign language poems which I was obliged to learn by rote over 40 years ago - but the problem with rote learning is that it relies excessively on memory and, in the absence of real understanding or integration, one is not in a position to be creative.

A much better approach is to build your understanding together with your knowledge, by continual reflection and integration of any new knowledge with one's existing knowledge. Continual reflection and introspection is needed to get proper integration: in those cases where one fails to achieve proper integration, it means that either one's previous conceptions or one's new conceptions are faulty and both must be revisited until they are reconciled. Failure to do this means that one's knowledge will be of a glib and superficial nature - somewhat like that of a confidence trickster.

To be more than a journeyman of science, one must be intellectually creative. This requires developing habits of thought that involve careful observation and questioning. If you ride a bicycle, for instance, you should observe that in order to turn right, you have to turn the handlebars to the left and you should ponder why this is. If you use a "cell 'phone" (called a "mobile 'phone" in some countries) you should question why it is so-called - what is the "cell" that is being referred to?

Properly developed, this way of thinking becomes habitual and it enables one to continually build and integrate one's knowledge - mostly going forward, incrementally, but sometimes going back to dismantle some misconception which may be familiar but which doesn't fit with some new insight. This way of thinking and learning requires introspection, retrospection, cogitation (thinking) and metacogitation (thinking about how you think) - and it is by developing these processes that one can learn to be creative.

Intellectual creativity essentially involves the ability to manipulate concepts in the abstract - to move them around and try new combinations. "What if" questions are a part of this process and so is having, or making, sufficient leisure time for abstract reflection. It is no accident that Archimedes had his "eureka" moment while taking a bath or that Newton achieved his great insights during a period of enforced inactivity, the universities in England having been closed due to the plague. In this way one can become a "life-long learner", to use a currently fashionable phrase. Rote learning is problematic in that it hinders development of the processes required for one to become a creative thinker. The challenge for teachers is to encourage creative thinking in a system governed by grades. Real education is much more than grades, though (Einstein, for example, achieved only mediocre grades), and is largely a process which you have to do for yourself. Although it is a painful process, which involves much work, I wish you success because there are few things more rewarding than a rich intellectual life.

*Clive Dennison*

# Chapter 1

# Basic physical concepts applicable to the isolation of proteins

This chapter is provided as a reference source to some of the basic physical concepts applicable to the isolation of proteins. Don't be put off by the mathematics in this chapter, or the fact that it seems somewhat removed from actual protein isolation. If you wish, you could skip this chapter and start at Chapter 2, coming back here when you need a definition or to gain greater clarity on an underlying physical concept.

**Linear motion**: Motion is important in protein isolation because, in order to separate molecules, at least some of them must be moved. Linear motion is a vector quantity, having magnitude and direction. The magnitude is the displacement: in the S.I. system the unit is the meter.

**Linear velocity**: Velocity $v$ is the displacement per unit time. In the S.I. system the unit is meters per second ($m.s^{-1}$).

$$v = \frac{dx}{dt}$$

Since displacement is a vector, velocity is also a vector.

**Acceleration**: An acceleration is a change in velocity in time. The change may be in either magnitude or direction. The unit of acceleration in the S.I. system is $m.s^{-1}.s^{-1}$, or $m.s^{-2}$. This precise definition of the word "acceleration" is not to be confused with the every-day meaning which is, roughly, "to go faster".

**Force**: A force is a 'push' or a 'pull' exerted on a body. It is a vector quantity, so it has magnitude and direction. In the S.I. system the unit of magnitude of a force is the newton (N).

**Newton's laws of motion.**

1. A body will maintain its state of rest or of uniform motion (i.e. at constant speed along a straight line) unless acted on by an unbalanced force. This is also called the law of inertia.
2. An unbalanced force $F$ acting on a body produces in it an acceleration $a$ in the direction of the force. The acceleration is directly proportional to the force and inversely proportional to the mass $m$ of the body.

   i.e.,
   $$a = k\frac{F}{m} \tag{1.1}$$

   where, $k$ is a proportionality constant.

   Eqn 1.1. defines the unit of **mass**, which refers to the *inertia* of a body, or its tendency to resist acceleration. In the S.I. system, the unit of mass is the kilogram which is that mass, when acted upon by a force of 1 newton, will have an acceleration of 1 $m.s^{-2}$. Hence, in the S.I. system of units, $k = 1$ and, so:

   $$F = ma \tag{1.2}$$

   | An important caveat applies to the 2nd law. To apply this law, an observer must be in an inertial frame, i.e. a non-accelerating reference frame in which the law of inertia (Newton's 1st law) applies. |
   |---|

3. If one object exerts a force on a second object, then the second object exerts an equal and opposite force on the first.

**Weight**: The *force* resulting from the effect of gravity acting upon a body (i.e. on a mass). Since weight is a force, in the S.I. system its units of magnitude are newtons.

As $F = ma$,

so, weight (newtons) = mass (kg) x $g$ ($m.s^{-2}$).

Where $g$ is the acceleration due to gravity (on Earth, $g$ averages 9.8 $m.s^{-2}$).

**Uniform circular motion**: Uniform circular motion is relevant to centrifugation, one of the standard tools in biochemistry. When a particle undergoes uniform circular motion, its speed in the direction of the tangent is the same at all points and its acceleration is also constant and acts along the radius towards the centre. The rate of rotation about the centre can be expressed as either tangential velocity or as angular velocity. Each expression has its own usefulness.

**Tangential velocity.** The tangential velocity $v_t$ is the instantaneous linear velocity ($dx/dt$) of a point at a radius $r$ from the centre of rotation, at a tangent to the circle.

**Angular displacement and angular velocity**. The angular displacement is the angle $\theta$ through which the radius to the particle has rotated from its initial direction.

$\theta$ may be expressed in **degrees** (°), **revolutions** (rev) or **radians** (rad). One revolution is equal to 360° or $2\pi$ rad. The radian is the S.I. unit for angular displacement.

$$\theta\,(\text{in radians}) = \frac{\text{arc length}}{\text{radius}}$$

Radians are thus the ratio of two lengths and consequently have no units.

The angular velocity $\omega$ is the angular displacement (in radians) per unit time, with units of rad $s^{-1}$.

$$\text{Angular velocity} = \frac{\text{angular displacement}}{\text{elapsed time}}$$

i.e.

$$\omega = \frac{d\theta}{dt} \quad \text{rad.s}^{-1}$$

The tangential velocity $v$ (m.$s^{-1}$) is related to the angular velocity $\omega$ (rad.$s^{-1}$), by the equation:-

$$v = \omega r$$

Another common expression of angular velocity is "revolutions per minute" (rpm).

**Centripetal acceleration and centripetal force.** Newton's first law states that the natural state of a body is to be at rest or in motion *at constant velocity in a straight line*. A body moving at constant tangential speed $v$, in a circular path with a radius $r$, is obviously not moving in a straight line - it is constantly changing direction and thus constantly accelerating, towards the centre of rotation as mentioned above. This is called centripetal acceleration.

"Peto" = "I seek" Hence "centripetal", seeking the centre

The magnitude of the centripetal acceleration $a_c$ is given by:-

$$a_c = \frac{v^2}{r}$$

Expressed in terms of angular velocity, the centripetal acceleration is:-

$$a_c = \omega^2 r$$

Such acceleration is the result of the application of a force, known as the centripetal force. Centripetal force $F_c$ - the force required to keep a body of mass $m$, moving on a circular path of radius $r$ at a tangential speed $v$ - is described by the equation:-

$$F_c = m a_c = \frac{mv^2}{r}$$

The centripetal force always points to the centre of the circle and thus continually changes direction as the body moves.

**The equivalence of gravity and acceleration.** A gravitational field is a location in space where a body (mass) experiences a force due to the gravitational attraction of another mass. This force is known as the weight of the body.

Consider two identical laboratories, one on Earth and one in space, far from any massive body, so that it experiences no significant gravity. A ball released by an experimenter in the laboratory on Earth, will be observed to accelerate towards the floor, with an acceleration of $g$, i.e. 9.8 $m.s^{-2}$ (*Figure* 1a). A similar ball, released in the laboratory in deep space, will simply remain where it is, because no force acts on it.

If the space laboratory was mounted on a rocket and accelerated at 9.8 $m.s^{-2}$, when the ball is released the observer will see it apparently accelerating towards the floor, at 9.8 $m.s^{-2}$, just as on Earth (*Figure* 1b). If the observer is unable to see outside, they would have no way of knowing whether they are on Earth, or are accelerating, in space. This is known as the equivalence of gravity and acceleration, and is one of the postulates of Einstein's general theory of relativity.

In the accelerating laboratory, the observer will attribute the apparent downward acceleration of the ball to some force. In reality, however, the ball is simply remaining where it is and it is the laboratory which is accelerating upwards. The apparent force accelerating the ball downwards is thus a fictitious force which is only evident to the accelerating observer[1]. The error which the observer makes is in applying Newton's 2nd law while in an accelerating frame, whereas Newton's 2nd law only applies to inertial frames, i.e. non-accelerating frames, where Newton's 1st law applies[1].

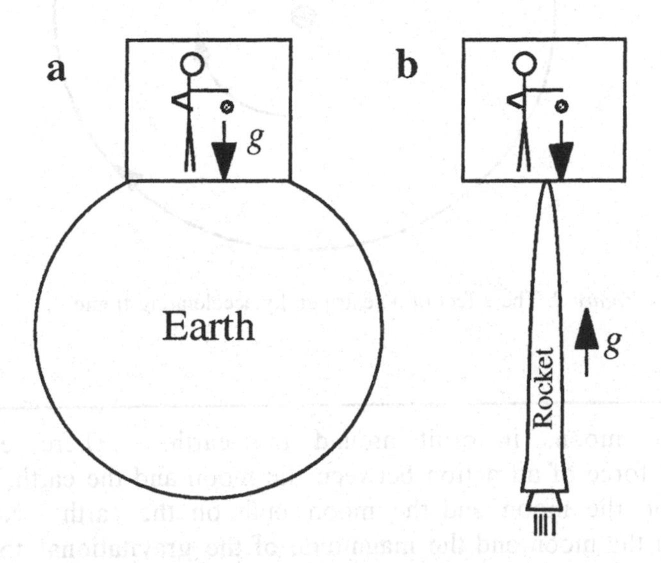


*Figure 1*. The equivalence of gravity and acceleration. a) In a laboratory on Earth, a released ball accelerates to the floor with an acceleration of $g$ (i.e. 9.8 $m.s^{-2}$). b) In a laboratory, accelerating at g, a released ball appears to an observer in the laboratory to behave exactly as on Earth

Now consider the laboratory to be mounted on a giant centrifuge and rotated such that it is given a centripetal acceleration of 9.8 $m.s^{-2}$. Imagine that this centrifuge is in deep space, away from any significant gravitational forces. Again, the observer will feel exactly as if they were on Earth and a released ball will appear to accelerate to the floor at 9.8 m.s$^{-2}$ (*Figure 2*). The force to which the centripetally-accelerating observer will attribute this apparent acceleration is again fictitious, but is known as the **centrifugal force**.

Centrifugal force is a fictitious force only apparent to an observer in a centripetally-accelerating frame.

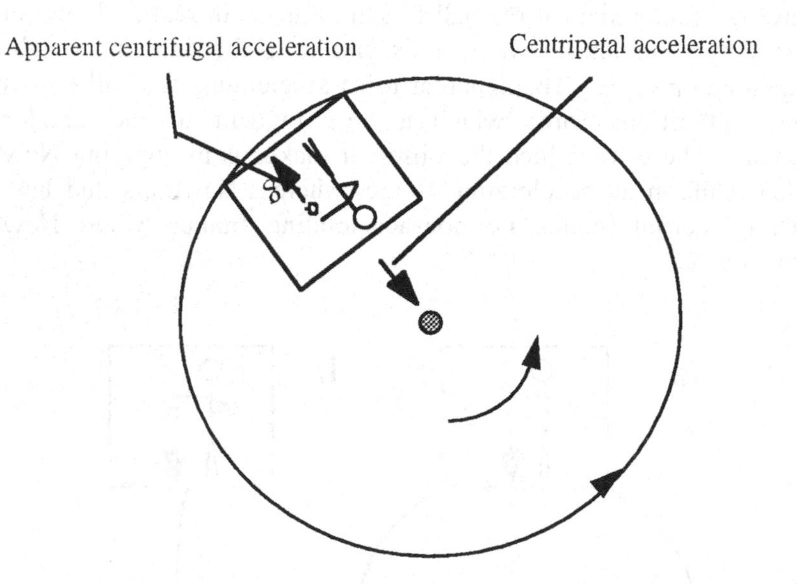


*Figure 2.* The effect of a centripetally accelerating frame

---

Consider the moon, in orbit around the earth. There exists a gravitational force of attraction between the moon and the earth, i.e. the earth pulls on the moon and the moon pulls on the earth. No other forces act on the moon and the magnitude of the gravitational force is a function only of the distance between the earth and the moon and their masses, i.e. it does not depend on whether the moon is moving or not. However, the moon **is** moving and the gravitational pull of the earth is just sufficient to give it the centripetal acceleration required to keep it in uniform circular motion. The only force which the moon exerts on the earth is an equal-but-opposite gravitational pull, i.e. there is no centrifugal force.

---

## Artificial gravity

Prolonged exposure to conditions of weightlessness have negative effects on human physiology, including weakening of bones and skeletal muscles. To counteract this, it is proposed that future space travellers should be exposed to artificial gravity, generated by rotating wheel-shaped space stations, so that the occupants are exposed to a centripetal acceleration, roughly equivalent to 9.8 $m.s^{-2}$, or *g*, the acceleration at the surface of the Earth, due to gravity. To the occupants of such a space station, conditions would appear to be exactly as on the surface of the earth.

We have an intuitive "feel" for our weight on Earth. If we stood on a planet where the gravitational force was twice that on Earth, i.e. where an unrestrained body would have an acceleration of 19.6 $m/s^2$, we would have a weight twice as much as our normal weight on Earth. Similarly, if we were given a centripetal acceleration of 19.6 $m/s^2$, we would have an apparent weight twice that on Earth.

It is common, therefore, to express accelerations in multiples of "*g*", the acceleration at the surface of the Earth, due to gravity. An acceleration of 19.6 $m/s^2$ could therefore be expressed as "2 x *g*". If we were placed in a giant centrifuge and given a centripetal acceleration of 6 x *g* (i.e. 58.8 $m/s^2$) we would feel six times heavier than normal.

However, although the effects may be the same, the accelerations on Earth and in a centrifuge have a different origin (force of gravity *vs* centripetal force) and so *there is no actual increase in gravitational force in a centrifuge*. In the following parts of this book, it will often be necessary to draw analogies between what happens in a gravitational field and what happens in a centrifuge. As a linguistic device, therefore, I will borrow the term "artificial gravity", from the field of space flight and will refer to conditions "under real or artificial gravity".

Similarly, the term "apparent weight" will be used to mean:-

$$\text{Apparent weight (newtons)} = \text{mass (kg)} \times g \text{ (metres/sec}^2\text{)}$$

where the acceleration is due to "real or artificial gravity", i.e. to gravity or to centripetal acceleration in a centrifuge.

**Buoyancy.** Buoyancy is a phenomenon associated with fluids and is a concept which crops up in many biochemical separation techniques, including salting-out, centrifugation and electrophoresis, so a sound understanding of the principles of buoyancy is necessary for a sound understanding of these separation techniques.

**Fluid**: In science a "fluid" is anything that can *flow*, i.e. it may be a liquid or a gas, but not a solid or a gel. This is a departure from every-day language, where the word "fluid" is synonymous with "liquid".

**Buoyancy force**: A buoyancy force acts upon a body immersed in a fluid, in real or artificial gravity, in a direction opposite to that of the weight of the body and is proportional to the *weight* of fluid displaced (This is a particular exposition of what is known as Archimedes' principle).

Note: i) the buoyancy force is *not* proportional to the mass or the volume of the fluid displaced (except insofar as these are related to the *weight* of fluid displaced).

ii) as the *weight* of the fluid displaced is a function of gravity, it follows that the buoyancy force will also be a function of the real or artificial gravity, i.e. the buoyancy force increases as the effective gravity increases.

**Buoyancy-driven fluid flow** is an outcome of the interplay of the forces of buoyancy and weight in fluids. If, in real or artificial gravity, a body of fluid displaces an equal volume of fluid of greater density, the weight of the displaced fluid will be greater than the weight of the displacing fluid and the latter will experience a buoyancy force greater than its weight and will thus be accelerated upwards (in every-day language, it will "float"). Conversely, if a body of fluid displaces a volume of fluid of lesser density, it will experience a buoyancy force less than its weight and it will thus be accelerated downwards (in every-day language, it will "sink").

There are many instances of buoyancy-driven fluid flow which are relevant in the techniques used for protein isolation, e.g. when proteins are concentrated during the stacking phase of disc electrophoresis, or during isoelectric focusing, as the protein is concentrated its density increases and, if not controlled, this would give rise to buoyancy-driven fluid currents.

One example, of buoyancy-driven fluid flow is **convection**. In physics, convection (the root of the word means "to convey") is considered as one of the ways in which heat may be transported (the others being conduction and radiation). When a body of fluid is heated, it (usually) expands and becomes less dense. It will consequently experience a buoyancy force which exceeds its weight and will thus rise. Conversely, if a body of fluid is cooled, it will usually sink (water, between 4°C and 0°C, is an exception). The only difference between convection and other forms of buoyancy-driven fluid flow, therefore, is that differential heating causes the differential densities which drive convection and, in consequence, heat is transported in the resulting fluid flow. Other forms of buoyancy-driven fluid flow may have nothing to do with heat.

For a buoyancy-driven fluid current to arise three criteria must be met (whether the process is convection or not):-

i) The medium must be fluid (since only fluids can flow to produce a current).
ii) The medium must exist in real or artificial gravity.
iii) A more dense part of the fluid must be above (or at the same height as) a less dense part of the fluid.

**Thrust**: A force which propels a body through a fluid medium.

**Drag**: A nett force which resists the movement of a body through a fluid medium and which operates in a direction opposite to that of the thrust. The nett drag is due to a number of contributing factors, the most important of which (for the present purposes) are the, i) frontal area, ii) shape and, iii) wetted area.

The "wetted area" is the total area in which the surface of the body is in contact with the fluid. In general, drag is proportional to the wetted area and to the frontal area. Also, a "streamlined" shape will have a lower drag than one which is not streamlined.

i.e. $$D = kA_wA_f$$

where D is the drag, $A_w$ is the wetted area and $A_f$ is the frontal area. The magnitude of the coefficient, *k*, depends on the shape of the body. At the speeds encountered in protein isolation, the drag increases linearly with the speed of the fluid flow past the body.

**Terminal velocity**: If the thrust is constant, the speed of a body will initially increase and, since the drag increases with speed, the drag will also increase until it becomes equal in magnitude to the thrust, at which point the speed will remain constant. The terminal velocity is thus the speed of a body, moving through a fluid medium, at which the thrust and the drag are of equal magnitude. A protein undergoing electrophoresis, for example, moves at its terminal velocity.

**Gel**: A gel consists of a large volume of liquid immobilised by a relatively small amount of a cross-linked polymer which has an affinity for the liquid. Gels cannot flow like normal fluids, but small solute molecules can migrate through gels, e.g. by diffusion or electrophoresis, as if they were fluids.

**Diffusion**: Solutes diffuse from areas of higher concentration to areas of lower concentration until, ultimately, the concentration becomes the same throughout the solution. Diffusion is described by Fick's law of diffusion, which states that the mass *m* of a solute that diffuses in a time *t* through a solvent contained in a channel of length *L* and cross-sectional area *A* is,

$$m = D\frac{A\Delta Ct}{L}$$

where $\Delta C$ is the solute concentration difference between the ends of the channel and *D* is the **diffusion coefficient**. The S.I. unit for the diffusion coefficient is $m^2s^{-1}$.

Diffusion occurs in liquids, gels, gases and in some solids.

**Dielectric constant**: The dielectric constant is defined by the equation describing Coulomb's law of electrostatic attraction, i.e.:-

$$F = \frac{Z_1 + Z_2}{distance.D}$$

Where $D$ is a property of the medium separating two charges, which influences the force between the them. The dielectric constant is related to the polarity of the medium. Thus $D$ for water (a polar solvent) is 80 and for benzene ($C_6H_6$) (a non-polar solvent), it is 2. Ionic interactions are thus stronger in less polar media.

References

1. Weidner, R. T. and Browne, M. E. (1985) in Physics. Allyn and Bacon Inc., Boston, p126-127.

## Chapter 1 study questions

1. What property of a body is measured by a spring scale?
2. What property of a body is measured by a balance?
3. Is the moon falling freely towards the earth?
4. Why does a ship float?
5. A ship weighs 100 metric tons. How much water will it displace?
6. A plimsoll line is a symbol on the side of a ship, consisting of a circle with a horizontal line drawn through it. Its purpose is to indicate the maximum safe load that the ship can carry. A ship is loaded in the Mediterranean Sea until its plimsoll line is level with the water line. Will the plimsoll line of this ship be above or below the water line if the ship was to enter the Great Lakes of America? (Ignore the effect of any fuel the ship might consume in getting to the Great Lakes).
7. Imagine a ship of 100 metric tons floating in water. Now, in a thought experiment, imagine that the whole ship-plus-water was placed in a giant centrifuge and spun at 2 x $g$ (i.e. twice the

gravitational field strength at the Earth's surface). i) What would now be the magnitude of the buoyancy force? ii) Would the ship float higher, lower or at the same level as before?

8. A sphere, immersed in water on Earth experiences a buoyancy force of 10 newtons. What buoyancy force would the same sphere experience if immersed in water aboard a spaceship in orbit around the Earth?

9. A sky-diver reaches terminal velocity when which two forces are in equilibrium?

10. What changes when a skydiver deploys their parachute and how does this affect their terminal velocity?

11. Consider two identical ships, one maximally laden and one unladen, both travelling at 5 knots. Which will require the greater force to keep it moving at 5 knots? Explain.

# Chapter 2

## An overview of protein isolation

Isolating a protein may be compared to playing a game of golf. In golf, the player is faced with a series of problems, each unique and yet similar to problems previously encountered. In facing each problem the player must analyse the situation and decide, from experience, which club is likely to give the best result in the given circumstances. Similarly, in attempting to isolate proteins, researchers face a series of similar-yet-unique problems. To solve these they must dip into their bags and select an appropriate technique. The purpose of this book is thus to fill the beginner's "golf bag" with techniques relevant to protein isolation, hopefully to improve their game.

Developing a protein isolation is also somewhat like finding a route up a mountainside. Different routes have to be explored and base-camps established at each stage. Occasionally it will be necessary to return to the base of the mountain for further supplies, and haul these up to the established camps, before the next stage can be attacked. A successful climb is always rewarding and if an efficient route is established, it may become a pass, opening the way to further discoveries.

### 2.1 Why do it?

This book is about the methods that biochemists use to isolate proteins, and so it may be asked, "why isolate proteins?" Looked at in one way, living organisms may be regarded as machines with features in common with the entities that we commonly think of as "machines". A typical machine is made of a number of parts which interact, transduce energy, and bring about some desired effect. Mechanical machines have moving parts, while electronic machines move electrons. "Engines" convert energy to mechanical motion. Internal combustion engines, for example, convert chemical energy to mechanical motion. Similarly, living organisms such as the human body are complex machines made up of many interacting systems. Proteins constitute the majority of the working parts of these systems and there are thus diverse reasons for isolating proteins, viz.;

- **To gain insight**. As with any mechanism, to study the way in which a living system works it is necessary to dismantle the machine and to isolate the component parts so that they may be studied, separately and in their interaction with other parts. In this way, structural and genetic information may acquired. The knowledge that is gained in this way may be put to practical use, for example, in the design of medicines, diagnostics, pesticides, or industrial processes.
- **For use in Medicine**. Many proteins may themselves be used as "medicines" to make up for losses or inadequate synthesis. Examples are hormones, such as insulin, which is used in the therapy of diabetes, and blood fractions, such as the so-called Factor VIII, which is used in the therapy of haemophilia. Other proteins may be used in medical diagnostics, an example being the enzymes glucose oxidase and peroxidase, which are used to measure glucose levels in biological fluids, such as blood and urine.
- **For use in Industry**. Many enzymes are used in industrial processes, especially where the materials being processed are of biological origin.

In every case a pure protein is desirable as impurities may either be misleading, dangerous or unproductive, respectively. Protein isolation is, therefore, a very common, almost central, procedure in biochemistry.

## 2.2 Properties of proteins that influence the methods used in their study

It must be appreciated that proteins have two properties which determine the overall approach to protein isolation and make this different from the approach used to isolate small natural molecules.

- Proteins are **labile**. As molecules go, proteins are relatively large and delicate and their shape is easily changed, a process called denaturation, which leads to loss of their biological activity. This means that only mild procedures can be used and techniques such as boiling and distillation, which are commonly used in organic chemistry, are thus *verboten*.
- Proteins are **similar** to one another. All proteins are composed of essentially the same amino acids and differ only in the proportions and sequence of their amino acids, and in the 3-D folding of the amino acid chains. Consequently processes with a high discriminating potential are needed to separate proteins.

The combined requirement for delicateness yet high discrimination means that, in a word, protein separation techniques have to be very **subtle**. Subtlety, in fact, is required of both techniques and of experimenters in biochemistry.

## 2.3 The conceptual basis of protein isolation

In a protein isolation one is endeavouring to purify a particular protein, from some biological (cellular) material, or from a bioproduct, since proteins are only synthesised by living systems. The objective is to separate the protein of interest from all non-protein material and all other proteins which occur in the same material. Removing the other proteins is the difficult part because, as noted above, all proteins are similar in their gross properties. In an ideal case, where one was able to remove the contaminating proteins, without any loss of the protein of interest, clearly the total amount of protein would decrease while the activity (which defines the particular protein of interest) would remain the same (*Figure 3*).

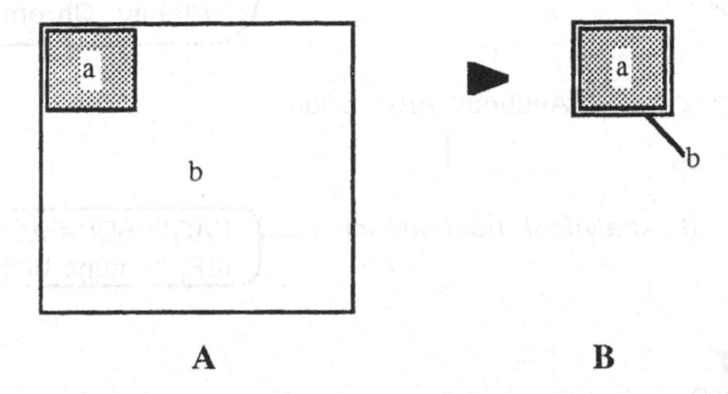


*Figure 3.* A schematic representation of a protein isolation.

Initially (*Figure 3*A) there is a small amount of the desired protein "a" and a large amount of total protein "b". In the course of the isolation, "b" is reduced and ultimately (*Figure 3*B) only "a" remains, at which point "a"="b". Ideally, the amount of "a" remains unchanged but, in practice, this is seldom achieved and less than 100% recovery of purified protein is usually obtained.

As a general principle, one should aim to achieve the isolation of a protein:-

- in as few steps as possible and,
- in as short a time as possible.

This minimises losses and the generation of isolation artefacts. Also, to further study the protein, the isolation will have to be done many times over and the effort put into devising a quick, simple, isolation procedure will be repaid many times over, in subsequent savings. The overall approach to the isolation of a protein is shown in *Figure 4*.

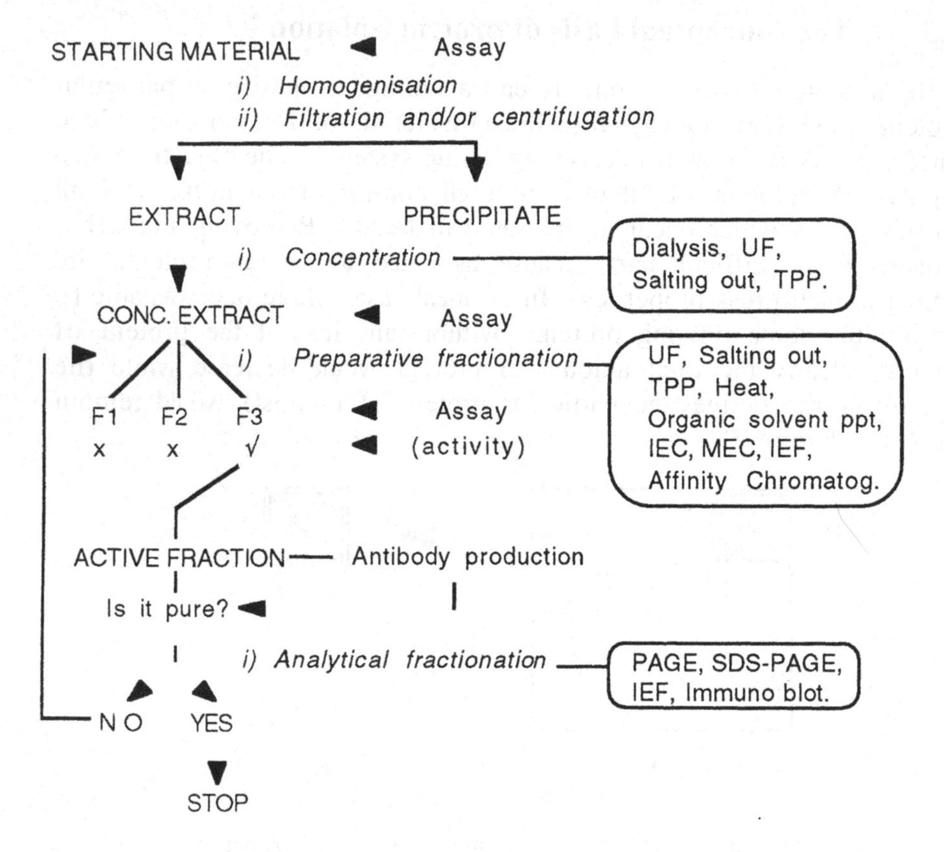


*Figure 4.* An overview of protein isolation.

### 2.3.1 Where to start?

To isolate a protein, one must start with some way of measuring the presence of the protein and of distinguishing it from all other proteins that might be present in the same material. This is achieved by a method which measures (assays) the unique activity of the protein. With such an assay, likely materials can be analysed in order to select one containing a large amount of the protein of interest, for use as the starting material.

The purpose of homogenisation is to get the proteins into solution where they may be manipulated.

Having selected a source material, it is necessary to extract the protein into a soluble form suitable for manipulation. This may be achieved by homogenising the material in a buffer of low osmotic strength (the low osmotic pressure helps to lyse cells and organelles), and clarifying the extract by filtration and/or centrifugation steps.

The clarified extract is typically subjected to preparative fractionation. If the protein of interest occurs in an organelle, then subcellular fractionation by differential centrifugation may be useful. It is necessary to assay the fractions obtained, in order to select the fraction(s) containing the protein of interest. The selected fraction(s) can then be subjected to further preparative fractionation, as required, until a pure protein is obtained.

Experience has shown that there is an optimal sequence in which preparative methods may be applied. As a first approach it is best to apply salting out (or TPP) early in the procedure, followed by ion-exchange or affinity chromatography. Salting out can, with advantage, be followed by hydrophobic interaction chromatography, because hydrophobic interactions are favoured by high salt concentration, so the necessity for desalting is obviated. The precipitate obtained from TPP, however, is low in salt and so can be applied directly to an ion-exchange system, without prior desalting. Generally, molecular exclusion chromatography should be reserved for late in the isolation when only a few components remain, since it is not a highly discriminating technique and it does not have a large sample capacity. Affinity chromatography often achieves the desirable aims of a rapid isolation using a minimum number of steps and so it should always be explored and preferentially used where possible.

Affinity chromatography should always be used as a first choice, where possible.

### 2.3.2 When to stop?

How can one know when the fraction is pure, i.e. when to stop? To obtain this information it is necessary to analyse the isolated fraction using a number of analytical fractionation methods. If a number of such analytical methods reveal the apparent presence of only one protein, it may be **inferred** that the protein is pure, and that the isolation has been successfully completed. Note, however, that it is not possible to **prove** that the protein is pure; one can merely fail to demonstrate the presence of impurities. Future, improved, analytical methods may reveal impurities that are not detected using current technology.

If, on the other hand, any analytical fractionation method demonstrates the presence of more than one protein, it may be inferred that the preparation is not pure. In this case, the application of further preparative fractionation methods may be required before the protein is finally purified.

As illustrated in *Figure 3*, the requirement is to remove as much contaminating protein as possible, while retaining as much as possible of the desired protein. Clearly then, to monitor the progress of an isolation, one needs two assays, one for the activity of the protein of interest (expressed in units of activity/ml) and another for the protein content (expressed as mg/ml). The activity per unit of protein (units/mg) gives a measure of the so-called *specific activity*. In the course of a successful protein isolation, the specific activity should increase with each step, reaching a maximum value when the protein is pure. It is also desirable that a maximum *yield* of the protein is obtained. The protein of interest is defined by its activity and so information concerning the yield may also be obtained from activity assays.

## 2.4 The purification table

The results of activity and protein assays, from a protein purification, are typically summarised in a so called purification table, of which Table 1 is an example.

*Table 1.* A typical enzyme purification table

| Step | Vol (ml) | Total protein (mg) | Total activity (units) | Specific activity (units/mg) | Purification (fold) | Yield (%) |
|---|---|---|---|---|---|---|
| Homogenate | 900 | 43600 | 48000 | 1.1 | (1) | (100) |
| pH 4.2 s'natant | 650 | 4760 | 28000 | 5.9 | 5 | 58 |
| $(NH_4)_2SO_4$ ppt | 140 | 1008 | 18667 | 18.5 | 17 | 39 |
| S-Sepharose | 57 | 7.1 | 7410 | 1044 | 949 | 15 |
| Sephadex G-75 | 35 | 2.45 | 3266 | 1333 | 1211 | 7 |

From an isolation of cathepsin L by R. N. Pike.

The figures in Table 1 are arrived at as follows:-

- *Volume* (ml) - this refers to the measured total solution volume at the particular stage in the isolation.
- *Total protein* (mg) - the primary measurement is of protein concentration, i.e. mg $ml^{-1}$, which is obtained using a protein assay.

Multiplying the protein concentration by the total volume gives the total protein (i.e. mg/ml x ml = mg).

- *Total activity* (units) - the activity, in units $ml^{-1}$, is obtained from an activity assay. Multiplying the activity by the total volume gives the total activity (i.e. units/ml x ml = units).
- *Specific activity* (units/mg) - the specific activity is obtained by dividing the total activity by the total protein. Alternatively, the activity (units/ml) can be divided by the protein concentration (mg/ml), in which case the "ml"s cancel out, leaving units/mg.
- *Purification* (fold) - "Fold" refers to the number of multiples of a starting value. In this case it refers to the increase in the specific activity, i.e. the purification is obtained by dividing the specific activity at any stage by the specific activity of the original homogenate. The purification "per step" can also be obtained by dividing the specific activity after that step by the specific activity of the material before that step.
- *Yield* (%) - the yield is based on the recovery of the activity after each step. The activity of the original homogenate is arbitrarily set at 100%. The yield (%) is calculated from the total activity (units) at each step divided by the total activity (units) in the homogenate, multiplied by 100. The yield can also be calculated on a "per step" basis by dividing the total activity after that step by the total activity before that step and multiplying by 100.

The efficiency of a step - is calculated as:-

$$\text{Purification (for that step)} \times \frac{\text{\% yield (for that step)}}{100}$$

## 2.5 Chapter 2 study questions

1. Why is protein isolation a common procedure in Biochemistry?
2. What distinguishes a protein isolation from the isolation of a small organic molecule?
3. What would one use as the starting material for the isolation of a particular protein?
4. In an ideal protein isolation, what is the yield of the desired protein?
5. Is such a yield ever achieved in practice?
6. If not, what yield should be aimed for?
7. Define the "specific activity" of a protein.
8. How does one know when to stop a protein isolation?

# Chapter 3

# Assay, extraction and subcellular fractionation

The object of extraction procedures is to get the proteins of interest out of the cellular material where they occur and into solution where they may be manipulated. The distribution of the protein(s) of interest is determined by an assay, so one of the first steps necessary is the development of a suitable assay.

Homogenisation disrupts the tissue and breaks open cells to release their contents, including cytoplasmic proteins and organelles. If the protein of interest occurs in an organelle, there may be some advantage in doing a subcellular fractionation, to isolate the organelle before releasing the proteins from the organelle.

Both assays and homogenisation require the use of buffers and so it seems appropriate to start this chapter with a discussion of buffers.

## 3.1 Buffers

Proteins have a pH dependent charge and many of the properties of proteins change with pH. Consequently, in working with proteins, it is important to control the pH. This is achieved by the use of buffers, and so at the outset it is important to have some insight into buffers, to know which buffer to use for any particular purpose, and how to make up the buffer.

Buffers are solutions of weak acids or bases and their salt(s), which resist changes in pH. Weak acids and bases are distinguished from strong acids and bases by their incomplete dissociation. In the case of a weak acid the dissociation is:-

$$\mathrm{HA} \rightleftharpoons \mathrm{H}^+ + \mathrm{A}^-$$

and the dissociation constant is:-

$$Ka = \frac{[\mathrm{H}^+][\mathrm{A}^-]}{[\mathrm{HA}]}$$

Now, $pKa = -\log Ka$

Thus, $$pKa = -\log\frac{[H^+][A^-]}{[HA]}$$

$$= -\log [H^+] - \log\frac{[A^-]}{[HA]}$$

$$= pH - \log\frac{[A^-]}{[HA]}$$

Hence, $$pH = pKa + \log\frac{[salt]}{[acid]} \qquad 3.1$$

For a weak base (e.g. Tris) the dissociation is:-

$$HB^+ \rightleftharpoons H^+ + B$$

Using similar arguments to those above, it can be shown that in this case,

$$pH = pKa + \log\frac{[base]}{[salt]} \qquad 3.2$$

Equations 2.1 and 2.2 are forms of the Henderson-Hasselbalch equation, which can be written in a general form as:-

$$pH = pKa + \log\frac{[basic\ species]}{[acidic\ species]} \qquad 3.3$$

From which it can be seen that, when [basic species] = [acidic species], then,

$$pH = pKa.$$

A simple monoprotic weak acid, such as acetic acid, yields a titration curve such as that shown schematically in *Figure 5*. It will be noticed that when pH = p*Ka*, the solution resists changes in pH, i.e. it functions best as a buffer in the range pH = p*Ka* ± 0.5.

$CH_3COOH$ is the acidic species in this buffer and $CH_3COO^-$ is the conjugate base. It may be observed that a solution of acetic acid itself ($CH_3COOH$) *will have* a pH less than the p*Ka* of acetic acid. Conversely, a solution containing only sodium acetate *will have* a pH greater than the p*Ka* of acetic acid. It is important to understand this point in order to appreciate how to make an acetate buffer using the approach described in Section 3.1.1.

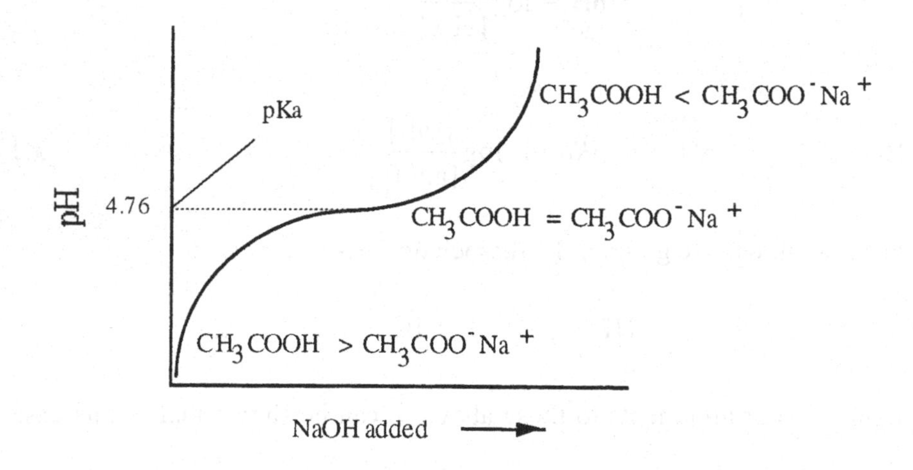


*Figure 5.* Schematic titration curve of a monoprotic acid, such as acetic acid.

A triprotic acid, such as phosphoric acid will yield a titration curve having three inflection points (*Figure 6*), corresponding to the three p*Ka* values of phosphoric acid.

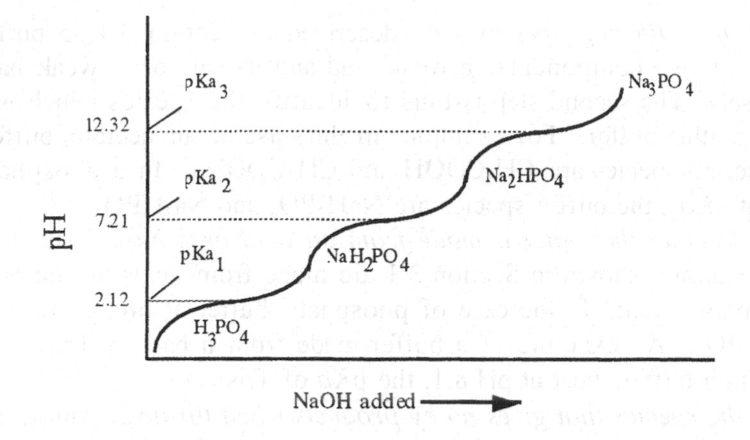


*Figure 6.* Schematic titration curve of phosphoric acid.

For most biochemical purposes, p$Ka_2$ is of greatest interest, since it is closest to the pH of the extracellular fluid of animals.

Note that:-

At p$Ka_2$, $[NaH_2PO_4] = [Na_2HPO_4]$.

At pH < p$Ka_2$, $[NaH_2PO_4] > [Na_2HPO_4]$.

At pH > p$Ka_2$, $[Na_2HPO_4] > [NaH_2PO_4]$.

Put another way, a solution of $NaH_2PO_4$ *will have* a pH less than p$Ka_2$ and a solution of $Na_2HPO_4$ *will have* a pH greater than p$Ka_2$. It is important to understand this point in order to appreciate how to make a phosphate buffer using the approach described below.

### 3.1.1 Making a buffer

A simple approach to the making of a buffer is described below. The advantage of this approach is that only *one* solution needs be made up. Several books suggest that buffers should be made up by adding "x" ml of a 1 M solution of "A" to "y" ml of a 1 M solution of "B". The problem with this approach is that it involves extra work (making up two solutions when one will do), waste (the unused volumes of "A" and "B" are discarded) and is usually inaccurate (the presence of extra salts and preservatives, for example, can change the pH due to common ion effects).

A simpler method follows the following steps[1]:-

- *Choose the buffer.* A buffer works best at its p*Ka* so the first step is to choose a buffer with a p*Ka* as close as possible to the desired pH.

- *Identify the buffering species.* As described in Section 3.1, a buffer consists of two components: a weak acid and its salt **or** a weak base and its salt. The second step is thus to identify the species which will constitute the buffer. For example, in the case of an acetate buffer, the buffering species are $CH_3COOH$ and $CH_3COONa$. In a phosphate buffer at p$Ka_2$, the buffer species are $NaH_2PO_4$ and $Na_2HPO_4$.
- *Identify whether the buffer is made from an acid or a base.* The two buffer examples shown in Section 3.1 are made from acids, acetic acid or phosphoric acid. In the case of phosphate buffer at p$Ka_2$, the acid is $NaH_2PO_4$. An example of a buffer made from a base is Tris/Tris-HCl, which buffers best at pH 8.1, the p*Ka* of Tris.
- *Choose the species that gives no by-products when titrated.* Almost all buffers can be made up by weighing out one component, dissolving in a volume just short of the final volume, titrating to the right pH, and making up to volume. It is **not necessary** to make up separate solutions of the two buffer constituents - the required salt can be generated *in situ* by titrating the acid with an appropriate base - or *vice versa* in the case of a buffer made from a base. [Remember: Titrate an acid "up" (i.e. with a strong base) and titrate a base "down" (i.e. with a strong acid)].

Remember, acid + base = salt + water

and, a buffer = (acid + its salt ) or (base + its salt).

The term "its salt" is important.
For example, if we wanted to make an acetate buffer, it is easy to identify that this buffer is made from acetic acid and its salt, say, sodium acetate. But,

**Q**: Could the required mixture of $CH_3COOH$ and $CH_3COONa$ be made by titrating a solution of $CH_3COONa$ to the correct pH with HCl?

**A**: No! Because the reaction in this case is:-

$$CH_3COONa + HCl \rightarrow CH_3COOH + NaCl$$

and the resultant solution contains NaCl, which is an unwanted by-product and which is not a salt of acetic acid (i.e. it is not "its salt").
On the other hand,

**Q**: Could the required mixture be made by titrating a solution of $CH_3COOH$ with NaOH?

**A**: Yes! The reaction in this case is:-

$$CH_3COOH + NaOH \rightarrow CH_3COONa + H_2O$$

which yields only the salt of acetic acid and water, i.e. there are essentially no by-products.
Similarly, in the case of a phosphate buffer, if one chooses $Na_2HPO_4$, the pH of a solution of this salt *will be* higher than p$Ka_2$ (see *Figure 6*) and this will require titration with an acid. If one chooses HCl, the reaction will be:-

$$Na_2HPO_4 + HCl \rightarrow NaH_2PO_4 + NaCl$$

which yields NaCl as an unwanted by-product. (And if one chooses $NaH_2PO_4$, this will change the phosphate molarity.) However, if one starts with $NaH_2PO_4$, the pH of a solution of this salt *will be* lower than p$Ka_2$ and this will require titration with a base. If one chooses NaOH, the reaction will be:-

$$NaH_2PO_4 + NaOH \rightarrow Na_2HPO_4 + H_2O$$

which yields only the desired salt ($Na_2HPO_4$) and water.
For a Tris buffer, one should start with the free base and titrate this with HCl to yield the salt of Tris, Tris-HCl.

- *Calculate the mass required to give the required molarity.* Having settled on the single buffer component to be weighed out, calculate the mass required to give the required molarity, when finally made up to volume. For example, the molarity of a phosphate buffer is determined by the molarity of the phosphate moiety, $HPO_4^{2-}$, which does not change when $NaH_2PO_4$ is titrated to $Na_2HPO_4$. If a litre of a 0.1 M buffer is required, then 0.1 moles of $NaH_2PO_4$ can be weighed out.
- *Add all other components, titrate and make up to volume.* Buffers often contain ingredients other than the two buffering species. For ion-exchange elution the buffer might contain extra NaCl, and buffers often contain preservatives such as $NaN_3$ or chelating agents such as EDTA. Except for $NaN_3$, these should all be added before the titration. All constituents should be dissolved in the same solution to just less than the final volume, i.e. a volume must be left for the titration but the final dilution after titration should be as small as possible. (The Henderson-Hasselbalch equation predicts that the pH of a buffer should not change with dilution, but this is only true over a small range, due to non-ideal behaviour of ions in solution.) Finally the solution is titrated to the desired pH and made up to volume. $NaN_3$ should be added after titration as it liberates the toxic gas, $HN_3$, when exposed to acid. Manganese salts should also be added after

adjustment of the pH as these may form irreversibly insoluble salts at pH extremes.

> $NaN_3$ should be added after titration as it liberates the toxic gas, $HN_3$, when exposed to acid.

### 3.1.2 Buffers of constant ionic strength

Besides pH, which influences the sign and magnitude of the charge on a protein, proteins are also influenced by the specific ions present in solution and by the solution ionic strength. In a buffer, the pH and the ionic strength are related. The Henderson-Hasselbalch equation, for a buffer made from an acid, is:-

$$\mathrm{pH} = \mathrm{p}Ka + \log\frac{[\mathrm{salt}]}{[\mathrm{acid}]}.$$

The ionic strength of the buffer is a function of the [salt]. Therefore, in this case as the pH rises, the buffer ionic strength also rises. Ionic strength is also a function of the molarity of the buffer. One can picture the relationship between the three variables, molarity (M), pH and ionic strength (I) as a lever, for which any one of the three could be fixed as a fulcrum and the relative movements of the other two observed (*Figure* 7).

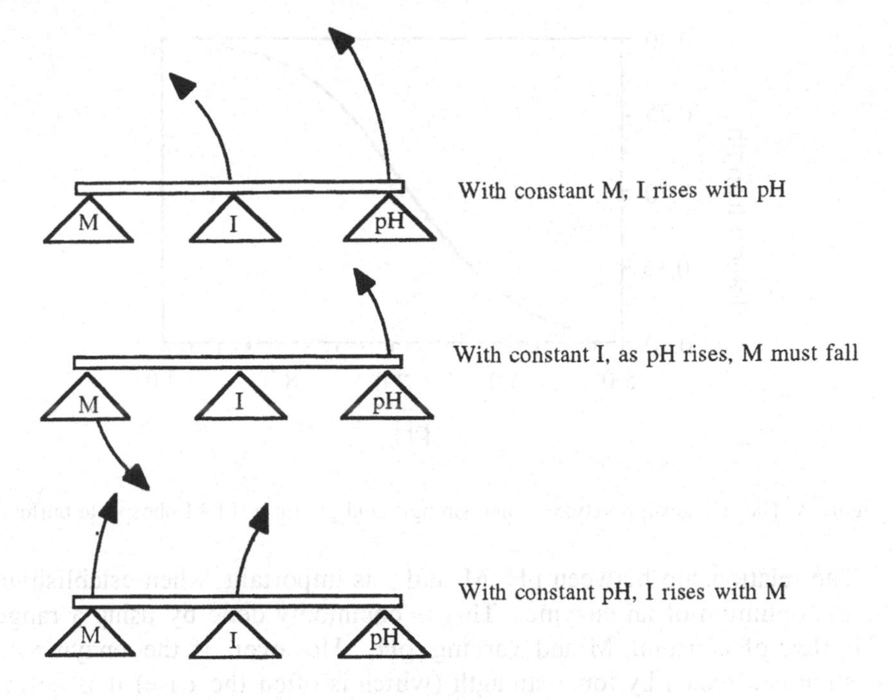
With constant M, I rises with pH
M
I
pH
With constant I, as pH rises, M must fall
M
I
pH
With constant pH, I rises with M
M
I
pH

*Figure 7.* The relationship between molarity, pH and ionic strength for a buffer made from a weak acid.

For a buffer made from a weak base, the relevant form of the Henderson-Hasselbalch equation is:-

$$\text{pH} = \text{p}Ka + \log\frac{[\text{base}]}{[\text{salt}]}.$$

In this case, therefore, the ionic strength increases as the pH decreases and the relationship between "M" (molarity), "I" (ionic strength) and pH can be visualised by reversing the positions of M and I in *Figure 7*.

The lever model shown in *Figure 7* must not be taken to imply a linear relationship between the variables. In fact, ionic strength changes sigmoidally with pH as shown in *Figure 8*. The 'rate' of change, i.e. $d$(ionic strength)/$d$(pH), is greatest at the p$Ka$, p$Ka_2$ in this case. The p$Ka$ itself also changes slightly with ionic strength[2,3]. The data in *Figure 8* were calculated according to Ellis and Morrison[2] (See p28).

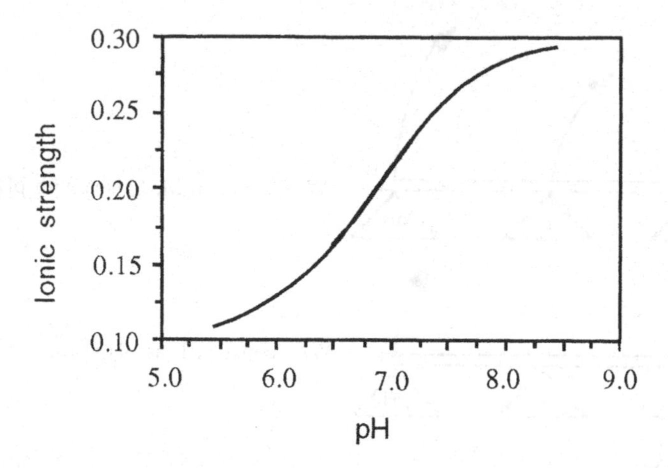


*Figure 8.* The relationship between ionic strength and pH for a 0.1 M phosphate buffer.

The relationship between pH, M and I is important when establishing the pH optimum of an enzyme. This is commonly done by using a range of buffers of constant M and varying pH. However, if the enzyme in question is affected by ionic strength (which is often the case) it is better to keep I constant and to vary M with pH (For an example, see ref. 4). The preparation of buffers of constant ionic strength is discussed by Ellis and Morrison[2]. An in depth discussion of buffers is provided by Perrin and Dempsey[5] and a table of p*Ka* values, useful in making buffers, is provided by Segel[6].

### 3.1.3 Calculating the ionic strength of a buffer (and the molarity of a buffer required to give a defined ionic strength)

The following MS BASIC program is an adaptation of that given by Ellis and Morrison[2]. Essentially the program calculates the ionic strength of the buffer of a given molarity and then determines what multiple of the answer is required to achieve the target ionic strength. In this approach, only buffer ions contribute to the ionic strength. An alternative is to use sodium chloride to make up the difference between the buffer ionic strength and the target ionic strength (assuming the former is smaller than the latter), as suggested by Ellis and Morrison[2]. When investigating the effects of ionic strength, it may be prudent to explore both approaches.

The example program given is for a 0.1 M phosphate buffer, but it may be easily adapted for other buffers by appropriate editing.

```
LPRINT "CALCULATION OF THE IONIC STRENGTH OF A BUFFER
OF GIVEN MOLARITY"
LPRINT "AND HENCE THE MOLARITY REQUIRED TO GIVE A
TARGET IONIC STRENGTH"

REM P1 = pKa of buffer
REM Z = Charge on conjugate base
REM IS = Ionic strength
REM C1 = buffer concentration
REM For phosphate buffer, P1 = 6.84, Z=-2
REM I = Target ionic strength
P1 = 6.84
Z=-2
C1=1
I=0.1
LPRINT "Buffer is phosphate"
LPRINT "pKa of buffer ="; P1
LPRINT "Charge on conjugate base ="; Z
LPRINT "Concentration of buffer ="; C1
LPRINT "Target ionic strength ="; I
REM H1 is lowest pH
REM H2 is highest pH
REM pH increment = X
H1 = 4!
H2 = 8!
X =.2
LPRINT "pH";CHR$(9) "Ionic strength"; CHR$(9) "Required
molarity"
LPRINT
FOR H=H1 TO H2 STEP X
X1=EXP((H-P1)*2.303)
A1=C1/(1+X1)
B1=A1*X1
U=((Z+1)*A1)+(Z*B1)
I1=.5*(((Z+1)*(Z+1))*A1+(Z*Z)*B1+ABS(U))
R=(I/I1)*C1
LPRINT H;CHR$(9)I1; CHR$(9)R
NEXT H
END
```

## 3.2 Assays for activity

Most proteins have some form of unique functional activity, which defines the specific protein and may be used to elaborate an assay for its detection and quantitation. A philosophical point to note is that it is necessary to conceive of an activity and to devise an appropriate assay, before the protein can be isolated. Ideally, the assay should be:-

- *specific*, to define the protein of interest and distinguish it from all others,
- *quantitative*, so that the success of the purification can be monitored, and,
- *economical* in terms of time and material.

The extent to which the assay meets these requirements has a major bearing on the difficulty, or otherwise, likely to be experienced in the subsequent protein isolation.

Assays for enzymes are usually specific although, for example, "proteolytic activity" may not be specific enough to be very useful by itself. On the other hand, an activity like "toxicity" may not be specific and may not be due to a single component. Since a large proportion of proteins isolated are enzymes, enzyme assays will be used to illustrate some of the conceptual dimensions of assays. It must be appreciated, however, that many proteins are not enzymes and different assay methods will be required for these.

### 3.2.1 Enzyme assays

Enzymes are biological catalysts which speed up the rate of specific reactions. The activity of an enzyme is therefore defined, and measured, by the extent to which it speeds up a reaction.

#### 3.2.1.1 The progress curve

The primary measurement in an enzyme assay is a progress curve, in which the amount of reaction that has taken place is plotted against time. The amount of reaction is defined as the amount of product formed or as the amount of substrate consumed. A typical progress curve, for an enzyme that is stable under the reaction conditions, is shown in *Figure 9*. The velocity of the reaction is given by the slope of the progress curve. Initially, the relationship between the amount of reaction and time is linear and the slope of this linear portion gives the initial velocity $V_o$. Eventually, the relationship becomes curvilinear and the reaction velocity (slope of the line) decreases, eventually reaching zero when the nett reaction stops. At this point, forward and reverse reactions are in equilibrium.

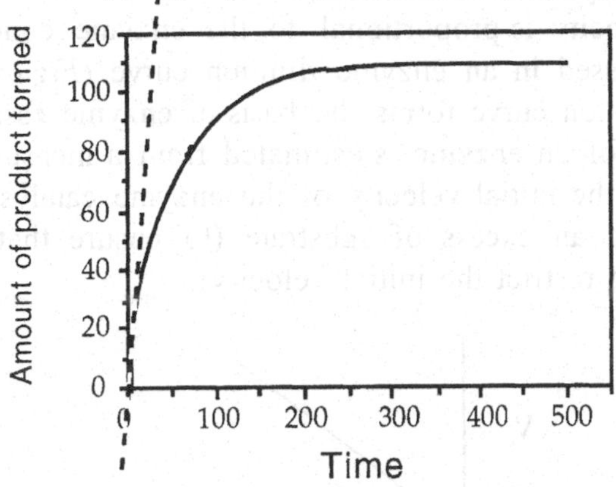


*Figure 9.* A progress curve for an enzyme-catalysed reaction.

The progress of an enzyme reaction may be visualised by considering the flow of water between two tanks, one initially empty and the other fairly full, with a pipe equipped with a tap connecting the two tanks at the bottom (*Figure 10*).

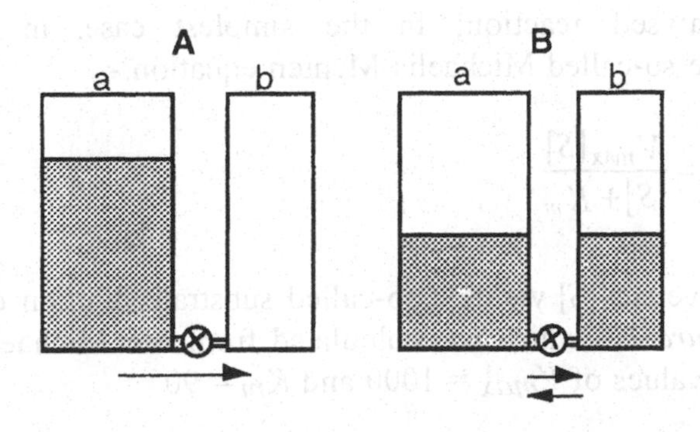


*Figure 10.* The water tank analogy of an enzyme-catalysed reaction.

In this analogy the volume of water in a tank is analogous to the concentration of reactant or product and the height (potential energy) is analogous to its chemical potential. Initially (A), there is a large amount of reactant (a) but no product (b). The reaction will therefore flow to the right until equilibrium is reached. The enzyme is equivalent to the tap in this model.

### 3.2.1.2 The enzyme dilution curve

The initial velocity is proportional to the enzyme concentration, a relationship expressed in an enzyme dilution curve (*Figure 11*). The linear enzyme dilution curve forms the basis of enzyme assays, in which the concentration of an enzyme is estimated from a measurement of its activity (i.e. from the initial velocity of the enzyme catalysed reaction), in the presence of an excess of substrate (to ensure that a substrate limitation does not restrict the initial velocity).

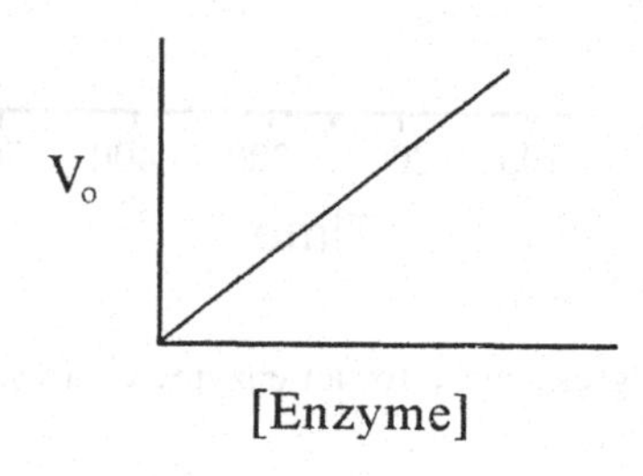


*Figure 11.* An enzyme dilution curve.

### 3.2.1.3 The substrate dilution curve

The concentration of substrate also affects the initial velocity $V_o$ of an enzyme-catalysed reaction; in the simplest case, in a manner expressed by the so-called Michaelis-Menten equation:-

$$V_o = \frac{V_{max}[S]}{[S]+K_m} \qquad 3.4$$

A plot of $V_o$ versus [*S*] yields a so-called substrate dilution curve, such as shown in *Figure 12*, which was calculated from the Michaelis-Menten equation, using values of $V_{max} = 1000$ and $K_m = 90$.

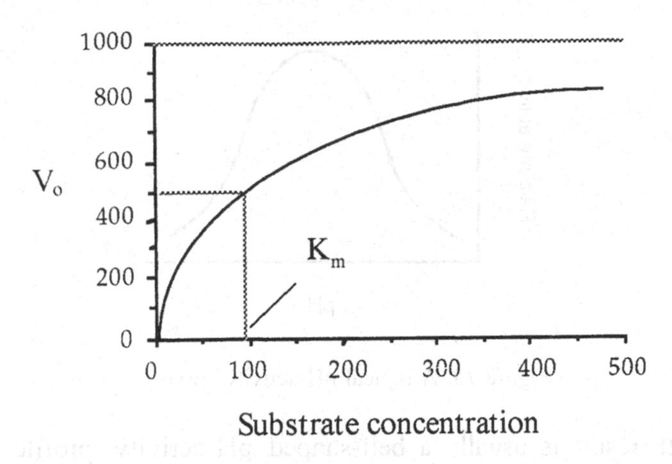


*Figure 12.* A substrate dilution curve.

**Note**: The substrate dilution curve must **not** be confused with the similarly-shaped progress curve.

The $K_m$, i.e. that substrate concentration which gives one half of the maximal velocity possible (at that enzyme concentration) is a constant, characteristic for a particular enzyme acting on a particular substrate. Knowledge of the $K_m$ is useful when devising an enzyme assay as it enables one to use a substrate concentration where $V_o$ will not be too sensitive to small changes in $[S]$ due to experimental error. A good rule-of-thumb is that $[S]$ should be as high as possible, preferably at a level where the substrate dilution curve is asymptotic to $V_{max}$. Often, however, $[S]$ is constrained by cost or experimental practicability, and values of less than $K_m$ may have to be used. For example the proteinase cathepsin B is routinely assayed at $[S] = \frac{1}{40}\ K_m$, using a fluorogenic substrate.

### 3.2.1.4 The effect of pH on enzyme activity

Another factor which influences $V_o$ is pH, which can exert its effect in different ways: on the ionisation of groups in the enzyme's active site, on the ionisation of groups in the substrate, or by affecting the conformation of the either the enzyme or the substrate. These effects are manifest in changes in the kinetic constants, $K_m$ and $k_{cat}$.

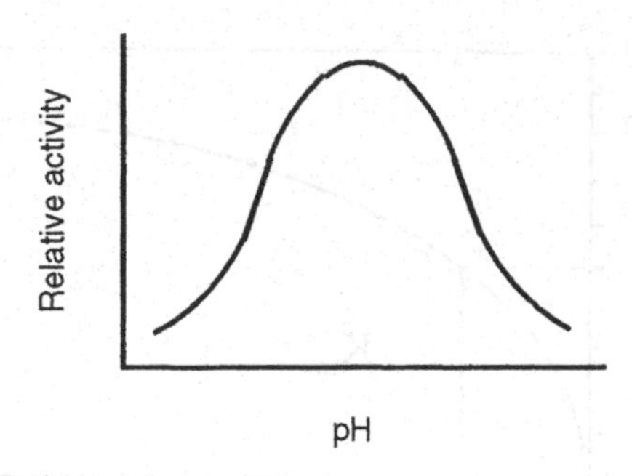


*Figure 13.* A typical pH-activity curve.

The nett result is usually a bell-shaped pH-activity profile (*Figure 13*). $V_o$ reaches its maximum at the optimum pH, which is the pH that should be used when assaying the enzyme. (See the discussion of pH vs ionic strength in Section 3.1.2.)

---

In expressing pH-activity profiles, many authors plot $k_{cat}/K_m$ against pH. Why, and what does this mean?.

For a reaction of the form:-

$$E + S = ES \rightarrow ES' \rightarrow E + P$$

the initial velocity, expressed as a function of the concentrations of **free** enzyme $[E]$ and substrate, is described by the equation:-

$$V_o = \frac{k_{cat}}{K_m}.[E].[S] \qquad 3.5$$

in which $k_{cat}/K_m$ is readily recognised as a second order rate constant. $k_{cat}/K_m$ is also known as the specificity constant as it is maximal with an optimum substrate.

Changes in pH will affect $V_o$, linearly, through effects on either (or both of) the enzyme's affinity for the substrate ($K_m$) or its turnover number ($k_{cat}$), but will not affect $[E]$ or $[S]$. The influence of pH is, therefore, essentially on $k_{cat}/K_m$ and $k_{cat}/K_m$ is maximal at the pH optimum.

The practical problem is that $[E]$, the concentration of *free* enzyme, is not known. In the measurements involved in establishing a pH-activity profile, the total enzyme concentration, $[E]_o$, and the initial substrate concentration, $[S]_o$, are constant (and known), and in the measurement of $V_o$ it can be assumed that $[S] = [S]_0$. The concentration of free enzyme, [E], is not known but is a function of [S] and $K_m$ as described by equation 2.6:-

$$[E] = \frac{[E]_o}{\frac{[S]}{K_m}+1} \qquad 3.6$$

$[E]$ is thus markedly influenced by the magnitude of $[S]$ relative to $K_m$. The variation of $[E]$ with $K_m$ is least when $[S]$ is small relative to $K_m$ (Modelling of equation 2.6 reveals that $[S]$ must be $\leq K_m/40$). When this is true (and *only* when this is true):-

- $[E] \approx [E]_0$.
- the shape of the pH-activity profile is linearly affected by changes in $k_{cat}$ and/or $K_m$, brought about by the changes in pH, and
- "relative activity" is proportional to $k_{cat}/K_m$.

When these conditions apply:-

$$V_o = \frac{k_{cat}}{K_m}[E]_0.[S] \qquad 3.7$$

which means that, if it is possible to use a substrate concentration $\leq K_m/40$, a pH-activity profile of $k_{cat}/K_m$ versus pH can be constructed from measurements of $V_o$ at different pH values and the known values of $[E]_0$ and $[S]$.

However, it is not always practicable to use a substrate concentration of $\leq K_m/40$ and when $[S]$ is *not* small relative to $K_m$, $[E] \neq [E]_0$ and the more familiar Michaelis-Menten equation applies, i.e.

$$V_o = \frac{k_{cat}}{K_m+[S]}.[E]_0.[S] \qquad 3.8$$

In this case separate measures of $k_{cat}$ and $K_m$ have to be obtained in the classical way by measuring $V_o$ at a number of levels of [S], at each pH. The paired data can be used to obtain estimates of $k_{cat}$ and $K_m$ at each pH, preferably by the method of Eisenthal and Cornish-Bowden[7]. From these, $k_{cat}/K_m$ values can be obtained and the pH-activity profile plotted.

---

#### 3.2.1.5 The effect of temperature on enzyme activity

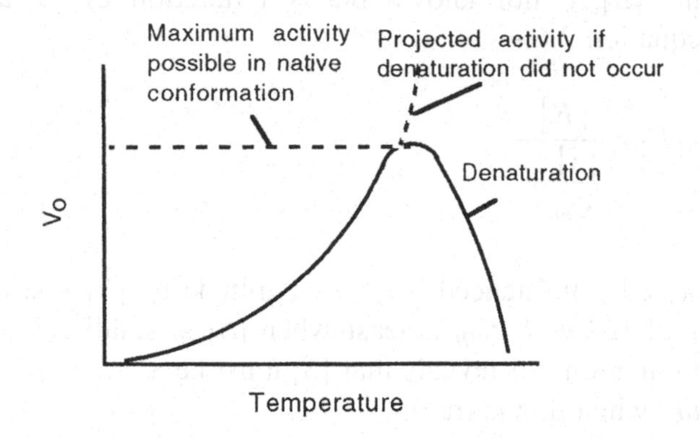


*Figure 14.* A typical temperature profile for an enzyme-catalysed reaction.

Finally, temperature also influences $V_o$. Two effects interact to give a resultant curve. On the one hand, like all chemical reactions, the velocity of enzyme-catalysed reactions increases with an increase in temperature, typically doubling for every 10°C rise in temperature. In the case of an enzyme-catalysed reaction, however, eventually a temperature is reached where the enzyme becomes unstable and begins to denature, at which point the reaction rate again declines. The resultant is usually an asymmetrical peak, which rises relatively slowly with an increase in temperature, and then drops rather suddenly (*Figure 14*).

It must be realised that denaturation is itself a reaction, with a temperature-dependent rate constant. Denaturation is generally a first-order reaction, since each protein molecule simply unfolds, independently of interaction with any other protein molecules. A useful way of expressing the temperature stability of an enzyme is therefore to measure the half-life ($t_{1/2}$) of its activity as a function of temperature. The "half-life" is the time taken for the enzyme activity to decrease from any value to half of that value. The half-life will be "infinite" until the temperature is reached at which the enzyme begins to denature. Thereafter, the half-life will decrease with an increase in temperature. The half-life of first-order reactions is discussed by Segel[6].

## 3.3 Assay for protein content

A number of methods are available for measuring protein concentration, each being based on a specific property of proteins, and each having certain advantages and disadvantages. Consequently, the different methods are more or less suitable for different applications and it is useful to have insight into these methods so that one can decide which one to use for a given application.

### 3.3.1 Absorption of ultraviolet light

UV-absorption is perhaps the most simple method for measuring the concentration of proteins in solution. A typical protein absorption spectrum has an absorption peak at 280 nm, due to the aromatic amino acids, such as tryptophan and tyrosine. Below 220 nm the absorption also increases strongly, due to peptide bonds, which absorb maximally at 185 nm. The extinction coefficients of different proteins tend to be different at 280 nm, due to their different aromatic amino acid contents, while below 220 nm the extinction coefficients are more similar. It is difficult to measure absorption accurately in this part of the spectrum, however, partly because oxygen forms begins to absorb in this region.

Because the extinction coefficients of proteins differ, UV-absorption is useful as a qualitative measure, for detecting the presence of protein, but is less useful for accurate quantitative measurements, except for pure proteins of known extinction coefficient. Because of its simplicity, UV-absorption is the method favoured for continuous (semi-quantitative) monitoring of the protein concentration in the eluate from chromatography columns.

One of the limitations of UV-absorbance, as a method for measuring protein, is that UV-absorbing, non-protein, compounds may interfere with the measurement. Nucleic acids, which are ubiquitously present in biological material, absorb UV radiation strongly, with a profile overlapping that of protein, but with a maximum at 260 nm. An elegant method for eliminating the absorption due to nucleic acids, thus allowing a measurement of protein in the presence of nucleic acid, has been proposed by Groves *et al.*[8].

In measuring the concentration of proteins by their UV-absorbance, remember that the extinction coefficient (or absorption coefficient) is given by the equation:-

$$A = a_m c l$$

where,

$A$ = absorbance
$a_m$ = molar extinction coefficient
$c$ = molar concentration of protein in solution
$l$ = length of the light path through the solution (usually 1 cm).

If the concentration is given in g/litre, then the equation becomes:-

$$A = a_s c l$$

where $a_s$ = specific extinction coefficient.
Note that $a_m$ = $a_s \times MW$.

### 3.3.2 The biuret assay

In alkaline solution, proteins reduce cupric ($Cu^{2+}$) ions to cuprous ($Cu^{1+}$) ions which react with peptide bonds to give a blue coloured complex. This reaction is called the biuret reaction and is named after the compound biuret (I), which is the simplest compound that yields the characteristic colour.

$$NH_2\text{-}\overset{\overset{\displaystyle O}{||}}{C}\text{-}NH\text{-}\overset{\overset{\displaystyle O}{||}}{C}\text{-}NH_2 \qquad \mathbf{I}$$

Because the reaction is with peptide bonds, there is little variation in the colour intensity given by different proteins. The biuret method can be used for the measurement of protein concentration in the presence of polyethylene glycol, a common protein precipitant.

A disadvantage of the biuret method is that it is relatively insensitive, so that large amounts of protein are required for the assay. A more sensitive variant of the method, the micro-biuret assay, has been devised[9], which overcomes this limitation to some extent. Another limitation is that amino buffers, such as Tris, which are commonly used in the pH range *ca.* 8-10, can interfere with the reaction.

### 3.3.3 The Lowry assay

The Lowry assay[10] may be considered as an extension of the biuret assay. Initially, a copper-protein complex is formed, as in the biuret assay. The cuprous ions then reduce the so-called Folin-Ciocalteu reagent[11], a phosphomolybdic-phosphotungstate complex, to yield an intense blue colour. An advantage of the Lowry over the biuret assay is that it is much more sensitive, and thus consumes much less of the protein sample. A disadvantage of the Lowry assay is that it is more sensitive to interference, a consequence of the more complicated chemistry involved. The Lowry assay has been reviewed by Peterson[12].

### 3.3.4 The bicinchoninic acid assay

Another development of the biuret reaction is the bicinchoninic acid (BCA) assay. Bicinchoninic acid forms a 2:1 complex with cuprous ions formed in the biuret reaction, resulting in a stable, highly coloured chromophore with an absorbance maximum at 562 nm[13,14]. The BCA assay is more sensitive than the Lowry method and is also less subject to interference by a number of commonly encountered substances. As the reaction is dependent, in the first instance, on the reduction of cupric ions to cuprous ions by the protein, it is sensitive to interference by strong reducing agents, e.g. ascorbic acid. This limitation also applies to the biuret and Lowry assays.

### 3.3.5 The Bradford assay

A protein assay which is rapidly becoming the most commonly used method, due to its simplicity, sensitivity and resistance to interference, is the dye-binding method described by Bradford[15]. Coomassie blue G-250, dissolved in acid solution, below pH 1, is a red-brown colour but regains its characteristic blue colour when it becomes bound to a protein. The concentration of protein can therefore be measured by the extent to which the blue colour, measured at 595 nm, is restored. Coomassie blue G-250 binds largely to basic and aromatic amino acids. Different proteins will differ in their content of these amino acids and so, ideally, a standard curve should be elaborated for each specific protein. A modification has been introduced by Read and Northcote[16] to overcome this problem to some extent. A disadvantage of the Bradford assay is that the reagent tends to stick to glass and plasticware. For this reason, the use of disposable cuvettes is recommended although, if necessary, the dye can be removed from surfaces by using SDS.

## 3.4 Methods for extraction of proteins

Once a promising source material has been identified using the activity assay described in Section 3.2, the next step is to extract the protein from this source. The objective in extracting proteins is to get them from the site where they occur in the tissue, into solution where they can be more easily manipulated and separated out. Most tissue proteins occur within cells, and possibly within organelles in the cells, and in these cases it is necessary to break open the cells and organelles, to release their protein contents. The methods chosen to disrupt the cells and organelles should be such that the proteins themselves are minimally damaged.

If the desired protein occurs within an organelle, then a useful purification of the protein may be achieved by a subcellular fractionation, whereby the different organelles are separated, before the protein is extracted from the organelle. Subcellular fractionation may be effected by differential centrifugation as described on p51.

Disruption of cells and organelles can lead to the formation of artefacts due to the release of enzymes and the mixing of enzymes and their potential substrates. Examples are proteolysis artefacts and the formation of artefactual thioester bonds. One way to minimise the formation of artefacts is to use 30% t-butanol in the homogenising buffer[17]. At 30%, t-butanol inhibits the activity of all enzymes tested to date. In most cases this inhibition is reversible, so the enzyme activity may be recovered by removing the t-butanol, either by dialysis or by TPP (See p80).

Homogenisation can lead to the formation of artefacts

### 3.4.1 Osmotic shock

A useful technique, which may be used in conjunction with mechanical means of disrupting cells, is the use of a buffer with a low osmotic pressure. In such a buffer water will tend to flow into the cells and organelles by osmosis, promoting their lysis and release of their proteins. To further promote the disruption of cell membranes, a low concentration of organic solvent, e.g. 2% n-butanol, is often added to the extraction buffer.

---

<u>**Laminar flow**</u>. A number of the homogenisers described below are dependent on the principle of the laminar flow of fluids for their operation. Laminar flow may be illustrated by taking a sheaf of paper sheets and throwing them onto a stationary surface. It will be observed that the bottom-most sheet of paper travels the smallest distance and the top-most sheet travels the greatest distance, due to the friction between the layers.

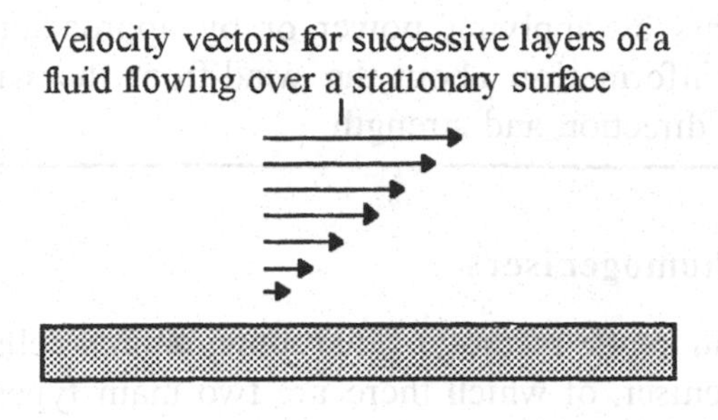

Velocity vectors for successive layers of a fluid flowing over a stationary surface


*Figure 15.* Laminar flow of a fluid.

Fluids (i.e. liquids or gases), with a viscosity greater than zero, flow over stationary surfaces in a similar way; the layer of fluid against the surface (the so-called boundary layer) is virtually stationary relative to the surface and successive layers travel at increasingly greater speeds.

An everyday example of the effects of laminar flow is the well-known phenomenon that one's voice can be heard to a greater distance downwind, than upwind. The speed of sound is about 1000 kph, which is high relative to common wind speeds, so the phenomenon is not due to the wind speed itself. Rather, the laminar flow of the wind distorts the sound waves, causing them to bend upwards into wind, and downwards downwind (*Figure 16*), so that the sound will be heard at greater distances, downwind.

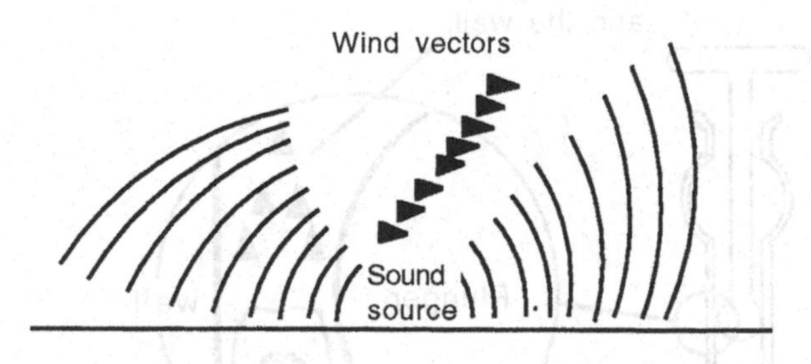


Wind vectors

Upwind, sound waves are bent away from the ground and can therefore only be heard a short distance from the source.
Downwind, sound waves are bent towards the ground and can therefore be heard further from the source.

*Figure 16.* The effect of the laminar flow of the wind upon sound waves.

Pilots of light aircraft with slow flying speeds, have to be especially conscious of the effects of laminar flow when landing. Landing is always done into wind to reduce the speed relative to the ground but, as the aircraft descends its airspeed will decrease and it may be necessary to compensate for this by applying power or by approaching with extra speed. Pilots get information about the wind from the windsock, which indicates the wind direction and strength.

---

### 3.4.2 Pestle homogenisers

An effective and gentle method of disrupting animal cells is by the use of a pestle homogeniser, of which there are two main types, Dounce and Potter-Elvehjem homogenisers. Pestle homogenisers generally disrupt whole cells but not organelles.

The Dounce homogeniser consists of a cylindrical glass tube, closed at one end, and two pestles (pistons) which fit into the cylinder with different clearances. Tissue is cut into small cubes, placed in the homogeniser with buffer and the "L" (loose) pestle is used first, to break the tissue into a fluid mixture. The "T" (tight) pestle is then used to disrupt the cells, releasing their contents. Typically, homogenisation is effected by a defined number of "passes" of the pestle, up and down the cylinder. Care should always be taken to support the end of the homogeniser against the bench, when it is being used, so that the end is not broken off by the hydraulic pressure within the cylinder.

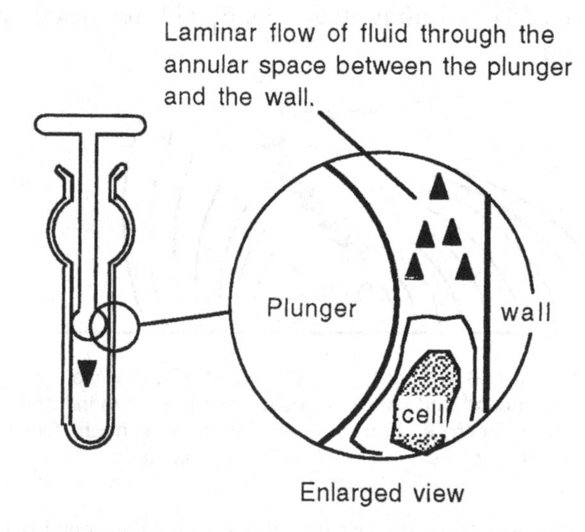


*Figure 17.* A Dounce homogeniser.

In a Dounce homogeniser, laminar flow of the fluid through the annular space between the pestle and the homogeniser wall results in different fluid speeds existing over the diameter of the cell, and the resulting shear forces disrupt the cell (*Figure 17*).

A Potter-Elvehjem homogeniser works in a similar way, except that the pestle has a more cylindrical shape, which induces shear forces over a greater area. Potter-Elvehjem homogenisers are available in automated, motorised versions.

### 3.4.3 The Waring blendor and Virtis homogeniser

These devices consist of a high speed stirrer with cutting blades, mounted in a glass vessel, the walls of which are indented from top to bottom, forming a clover-leaf cross section. The speed of the blades' motion generates strong shear forces, due to laminar flow, while the irregular outline of the vessel gives good overall mixing of the solution. The degree of disruption depends upon the speed of rotation of the blades. At high speeds, a blendor will disrupt mitochondria and nuclei and may even denature proteins. It is mostly used with plant and animal tissues but is less effective with micro-organisms.

Note that although it is a "blender", the trade name is "Waring blendor".

### 3.4.4 The Polytron/Ultra-Turrax-type homogeniser

"Polytron" and "Ultra-Turrax" are trade names for a type of homogeniser which consists of a stationary vertical tube, equipped with serrated teeth and radially distributed holes at its lower edge. Fitting closely into the stationary tube is a motor-driven tube, also with radially distributed holes corresponding to those on the stationary tube.

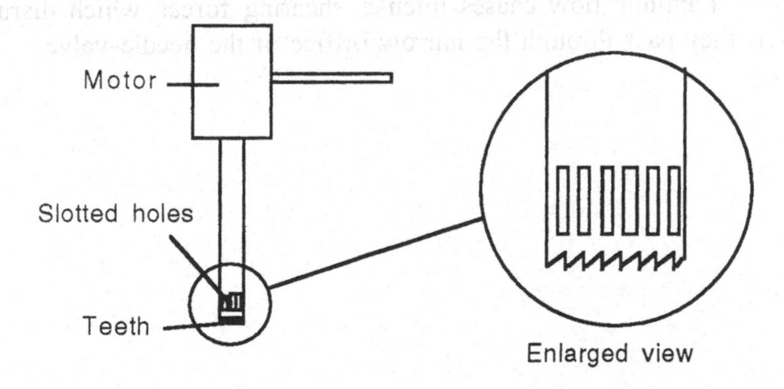


*Figure 18.* The Polytron/Ultra-Turrax type homogeniser.

Rotation of the inner tube causes the sample to be flung outwards, through the holes in the tubes. Because the two sets of holes continuously and rapidly come into and out of register, the sample gets chopped into small pieces and simultaneously homogenised by shear between the rotating and stationary tubes. Such homogenisers are very effective and only a short period of homogenisation is required, the sample being cooled in an ice bath during this period.

### 3.4.5 Grinding

Several types of apparatus are available for grinding. In the Edmund-Bühler disintegrator, bacterial cells are vibrated with glass beads in a jacketed container. Cells are broken by impact, tearing and maceration between the hard surfaces. To avoid heating, cooling liquid is circulated through the jacket.

### 3.4.6 The Parr bomb

In the Parr bomb, the sample is subjected to nitrogen gas under very high pressure. Under these conditions, the nitrogen dissolves in the cell fluids. When the pressure is released, the explosive generation of nitrogen bubbles causes disruption of the cell, and less frequently of organelles.

### 3.4.7 Extrusion under high pressure

In an apparatus such as the French pressure cell (*Figure 19*), cells are broken by extrusion through a narrow orifice at pressures of up to 8,000 p.s.i. Laminar flow causes intense shearing forces which disrupt the cells as they pass through the narrow orifice of the needle-valve.

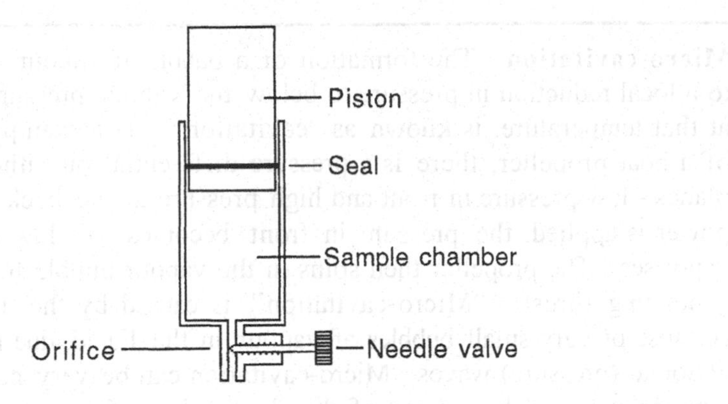


*Figure 19.* A French pressure cell.

---

**Needle valves**. Needle-valves are devices used to adjust or regulate the flow of fluids. They consist of a tapering "needle", with a round cross-section, which fits into a corresponding round hole, called a "jet". When the needle is retracted slightly from the jet, an annular gap is formed, between the needle and the jet, and fluid can flow through this gap. The cross-sectional area of the annular gap can be altered by adjusting the degree to which the needle is retracted from the jet.

An every-day application of needle-valves is in carburettors, where they are used to control the flow of petrol and air, to ensure a correct mixture of the two. Adjusting the needle-valves is one of the steps involved in "tuning" a carburettor.

---

### 3.4.8 Sonication

Application of high frequency sound waves is an effective method of cell breakage which can be applied to micro-organisms. The mechanism is thought to involve "micro-cavitation", i.e. the production of very local transient pressure differences, which break cell walls. The efficiency of cell breakage is influenced by the power output of the instrument, the duration of exposure and the volume of material processed. In general, the volume which can be treated in a given time is not great - not as great, for example as that using high pressure extrusion. Cooling is necessary to prevent the build-up of heat.

**Micro-cavitation**. The formation of a bubble of vapour in a liquid, due to a local reduction in pressure to below the vapour pressure of the liquid at that temperature, is known as "cavitation". For example, in the case of a boat propeller, there is a pressure differential on either side of the blades - low pressure in front and high pressure at the back. If too much power is applied, the pressure in front becomes too low and the water vaporises. The propeller then spins in the vapour bubble formed, without generating thrust. "Micro-cavitation" is caused by the formation and collapse of very small bubbles of vapour in the liquid, due to the passage of sound (pressure) waves. Micro-cavitation can be very corrosive and is a major cause of the erosion of ships' propellers, for example.
(See p63, "How hard can one "suck" on water")

### 3.4.9 Enzymic digestion

Enzymes provide a very gentle and specific means of disrupting cells to release their contents. For example, the cell walls of Gram-positive bacteria may be digested with the enzyme, lysozyme. Similarly plant cell walls may be digested with cellulases and fungal cell walls with chitinases.

## 3.5 Clarification of the extract

The cellular extract prepared by one of the methods described above may be clarified, by filtration through a nylon mesh or cheese-cloth, to remove the larger tissue debris, and centrifuged at relatively low speed to remove insoluble cell components.

## 3.6 Centrifugal subcellular fractionation

### 3.6.1 The centrifuge

A centrifuge is an instrument in which liquid samples are spun (centrifuged) about a vertical axis. The samples are contained in plastic or glass vessels which are placed in a rotor which is attached to a vertical shaft, driven by a motor. Rotation of the rotor exposes the samples to a centripetal acceleration and a consequent artificial gravity which is usually expressed in multiples of the acceleration due to gravity on Earth (see p7).

### 3.6.2 Principles of centrifugation

As an introduction to the topic of centrifugation and to gain insight into the forces acting upon a molecule undergoing centrifugation, it is useful to consider the derivation of the so-called **Svedberg equation**. Before reading this section, it might be useful to revisit the sections on uniform circular motion, artificial gravity and buoyancy, in Chapter 1. Newton's 1st law of motion states that any body in motion at a constant speed in a straight line will continue in that motion unless acted upon by a force. This force will cause the body to accelerate in the direction of the force, according to the equation:-

$$\text{Force} = \text{mass} \times \text{acceleration}$$

i.e.

$$F = ma$$

Consider a stone swung around on a piece of string at constant speed. Nothing goes any faster with time, so is the stone accelerating? Yes, because it is constantly changing direction. According to Newton's law, it requires a force to make a moving mass change direction and the force will make the mass accelerate. In this case the force is applied by the string and the direction of the force is always towards the centre of rotation and hence is known as a centripetal force.

As passengers in motor cars we have probably all experienced rotary motion and the consequent centripetal force (and the apparent centrifugal force - apparent to passengers in the centripetally accelerating frame of the car - see p6).

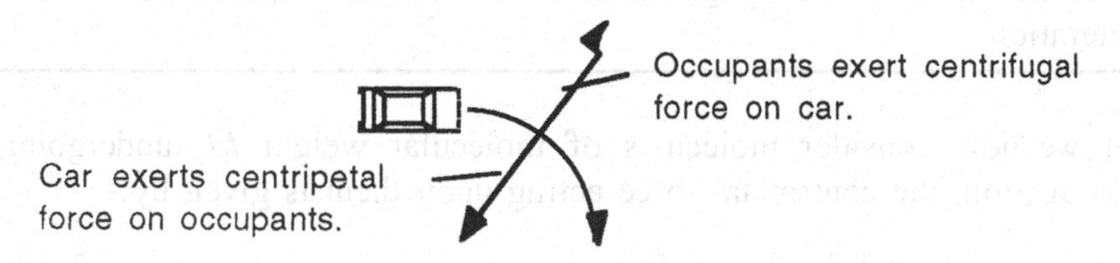
Occupants exert centrifugal force on car.
Car exerts centripetal force on occupants.

*Figure 20.* The equal and opposite forces acting in a cornering car.

When a car takes a corner, the occupants experience outwardly-directed artificial gravity, due to their centripetal acceleration $a_c$, described by the equation:-

$$a_c = \omega^2 r$$

where, $\omega$ = the angular velocity in rad·s$^{-1}$ (remember one revolution = $2\pi$ radians)

$r$ = radial distance from axis of rotation.

Since they are in an accelerating frame, the occupants will experience an apparent centrifugal force (see pp6-8). The centripetal force $F$ which the car exerts on the occupants (and the equal-but-opposite apparent centrifugal force which the occupants exert on the car) is a function of their mass $M$.

i.e. $$F=M\omega^2r \quad 3.9$$

---

From Eqn 3.9 we can see why a car will skid if it is turned too rapidly;

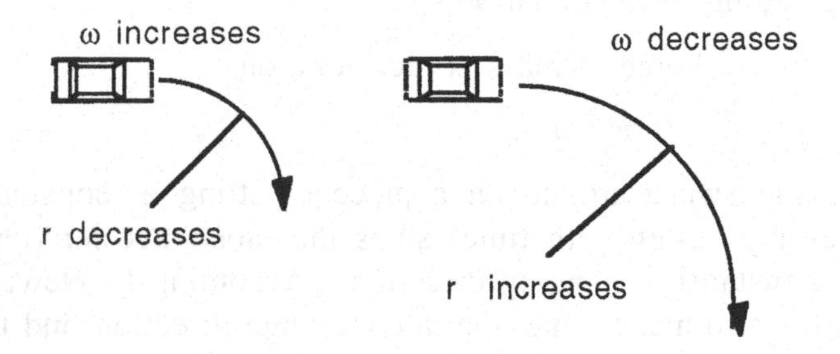


*Figure 21.* The effect of increasing the sharpness of a turn.

Increasing the sharpness of a turn increases $\omega$, but decreases $r$. In Eqn 3.9, however, $\omega$ is squared, while $r$ is linear, so the increase in $\omega$ has the greater effect. The centripetal acceleration is thus increased in a sharp turn and the car is more likely to skid, which it will do when the frictional grip of the tyres is no longer able to effect the required centripetal acceleration.

---

If we now consider molecules of molecular weight $M$, undergoing centrifugation, the centripetal force acting upon them is given by;-

$$\text{Centripetal force} = M\omega^2r \quad 3.10$$

The equal-but-opposite apparent centrifugal force causes the molecules to sediment down the centrifuge tube. As they start to move, however, they encounter a frictional resistance to their movement, given by:-

$$\text{Frictional force} = f\frac{dr}{dt} \quad 3.11$$

Where $f$ = frictional coefficient
$dr/dt$ = the rate of change in radius with time $t$.

The sedimenting molecule must also displace the solvent into which it sediments and this gives rise to a buoyant force which increases, with increasing $r$, as the molecule sediments:-

$$\text{Buoyant force} = M\omega^2 r v \rho \qquad 3.12$$

Where $v$ = partial specific volume of the molecules ($cm^3$ volume increase caused by 1g of solute),
$\rho$ = density of the solution.

---

The buoyancy force increases with the radius in the same way as the artificial gravity does. This is because the effective weight of the solution is a product of its mass and gravity (see p7) so, as it sediments, the molecule will displace an increasing effective *weight* of the solution. To grasp what this means, conduct the following thought experiment:- Imagine a little ship floating on the surface of the water in a centrifuge tube, undergoing centrifugation, with its plimsoll line (see p11) on the water line. If $r$ is increased, by removing some of the water, the effective weight of the ship ($M\omega^2 r$) would increase but the ship would not float lower in the water because the effective weight of the displaced water and the consequent buoyancy force ($M\omega^2 r v \rho$) would increase by the same amount (*Figure 22*).

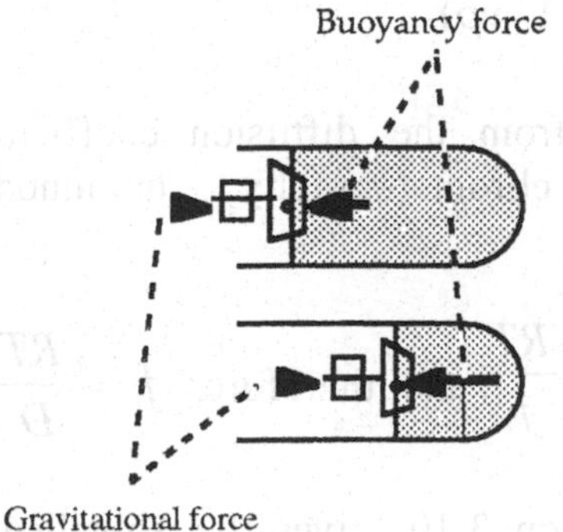


*Figure 22.* The effect of radius of rotation on effective weight and buoyancy forces.

---

Eqns 3.10 to 3.12 can be combined in the expression:-

$$\text{Centrifugal force} = \text{Buoyant force} + \text{Frictional force}$$

i.e. $$M\omega^2 r = M\omega^2 r\, v\rho + f\frac{dr}{dt}$$

Hence, $$M = \frac{f}{(1-\upsilon\rho)}\cdot\frac{\frac{dr}{dt}}{\omega^2 r} \qquad 3.13$$

i.e. if all other factors are kept constant, the rate of sedimentation depends upon the molecular weight.

A new term, "s", the sedimentation coefficient, can be defined (where "s" is a small capital letter):-

$$s = \frac{\frac{dr}{dt}}{\omega^2 r}$$

[i.e. it is the rate of sedimentation per unit of centripetal acceleration (which may be expressed as $m.s^{-2}$ or in multiples of $g$)]

Hence, $$M = \frac{f\,s}{(1 - v\rho)} \qquad 3.14$$

$f$ can be determined from the diffusion coefficient $D$, measured by molecular exclusion chromatography, immunodiffusion or ultra-centrifugation:-

$$D = \frac{RT}{f} \text{ and, therefore, } f = \frac{RT}{D}.$$

Substituting into equation 3.10, gives the so-called *Svedberg equation* named in honour of Professor The Svedberg, of Sweden:-

$$M = \frac{RTs}{D(1-v\rho)} \qquad 3.15$$

*Note*: "s" has the range $1 \times 10^{-13}$ to $500 \times 10^{-13}$. A new unit, "the Svedberg", abbreviated "S" (a capital letter), can be defined to obviate the ($\times 10^{-13}$).

i.e. $S = s \times 10^{13}$

### 3.6.3 Sub-cellular fractionation

Fractionation of sub-cellular organelles may be effected by centrifugation[18]. Assuming one is dealing with rigid, spherical particles, the time required to sediment a specific particle, subjected to centrifugation, is given by the equation[19];-

$$t = \frac{9}{2} \frac{\eta}{w^2 r_p^2 (\rho - \rho_0)} \ln \frac{R_m}{R_b} \qquad 3.16$$

where,
- $t$ = time required for particle to sediment from meniscus to bottom of the tube,
- $\eta$ = viscosity of the medium,
- $\omega$ = angular velocity,
- $r_p$ = radius of sedimenting particle,
- $\rho$ = density of sedimenting particle,
- $\rho_o$ = density of medium,
- $R_m$ = radius from centre of rotation to solution meniscus,
- $R_b$ = radius from centre of rotation to the bottom of the tube.

If the experimental set-up is established and the angular velocity is kept constant, Eqn 3.16 can be simplified to:-

$$t = K \frac{1}{r_p^2 (\rho - \rho_o)} \qquad 3.17$$

---

Note that if the angular velocity is **not** constant, Eqn 3.17 must be modified to:-

$$t = K'' \frac{1}{\omega^2 r_p^2} \qquad 3.18$$

So, for a given particle in a given system:-

$$t\omega^2 = \frac{K''}{r_p^2} \qquad 3.19$$

Hence, if we wish to change the time of sedimentation, the term $(t\omega^2)$ must be kept constant,

i.e. specified time x (specified rpm)$^2$ = new time x (new rpm)$^2$.

---

In the case of a sub-cellular fractionation, most particles are of about the same density and so equation 3.17 may be reduced to:-

$$t = K' \frac{1}{r_p^2} \qquad 3.20$$

i.e. the time taken to sediment a particle is inversely related to the square of its radius. This fact underlies the technique of fractionation by differential centrifugation, such as the example shown in *Figure 23*.

In a differential centrifugation fractionation, the "pellet" is unlikely to be pure because, initially, all particles of the homogenate are distributed evenly throughout the centrifuge tube. Upon sedimentation, the heaviest particles sediment first but other less dense material is dragged along and that originally near the bottom of the tube is co-precipitated. Repeated "washings" can improve the purity.

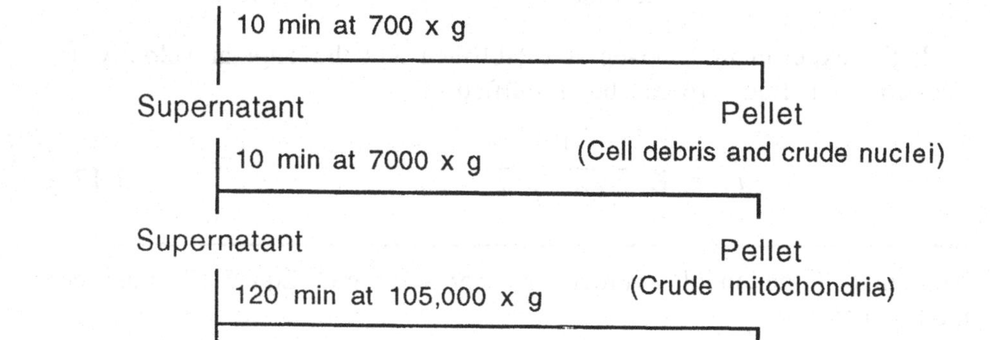


*Figure 23.* An example of sub-cellular fractionation by differential centrifugation.

### 3.6.4 Density gradient centrifugation

Re-consider equation 3.17, i.e.:-

$$t = K \frac{1}{r_p^2 (\rho - \rho_o)} \tag{3.17}$$

| | | | |
|---|---|---|---|
| If, | $(\rho - \rho_o) = 0,$ | $t = \infty$ | and particle will not sediment, |
| if | $(\rho - \rho_o) < 0,$ | $t < 0$ | and particle will have negative sedimentation in real time, i.e. it will rise to the top, |
| and if | $(\rho - \rho_o) > 0,$ | $t > 0$ | and particle will sediment. |

In each case, the behaviour of the molecule is determined by its density ($\rho$), relative to that of the medium ($\rho_o$). In real or artificial gravity, if the molecule is less dense than the medium, it will experience a positive buoyancy force and will consequently rise (e.g. as cream rises to the top of milk). Conversely, if it is more dense than the medium, it will sink. If it is the same density as the medium, it will experience a zero buoyancy force and will neither rise nor sink.

Using this information, one of two strategies for isopycnic (equal density) centrifugation can be adopted to separate particles by virtue of their differences in density:-

- The sample can be centrifuged in a solution having a density equal to that of the sample protein of interest. Proteins that are more dense or less dense will sediment or float to the top of the solution, respectively, leaving the protein of interest in solution. A difficulty with this approach, however, is that often the density of the protein of interest is not known beforehand.
- The sample can be layered on top of a continuous density gradient and centrifuged. If the gradient extends to densities exceeding the density of the sample proteins, these will sediment through the gradient until they reach points where their density is equal to that in the gradient, at which point sedimentation ceases, i.e. the proteins become focused at their isopycnic points.

Density gradients may be generated using sucrose dissolved in buffer.

Disadvantages of sucrose are:-

- It interferes with the Lowry and Bradford protein assays, though less with the latter.
- It can penetrate cells. This, of course, is a problem which only applies to the fractionation of cells.

Advantages of sucrose are:-

- It is biologically inert.
- It is inexpensive.
- It is dialysable and is thus easy to separate from the sample proteins.

"Ficoll", a Pharmacia product, consists of sucrose crosslinked with epichlorhydrin, and has a molecular weight of about 400,000. Like sucrose it is biologically inert, but in some respects it is opposite to sucrose, e.g. it cannot penetrate cells and so is suitable for fractionation of cells, but it is not dialysable and so is difficult to separate from proteins.

With sucrose or Ficoll, a density gradient can be generated using a two chamber gradient generator, with an insert in the low density chamber to compensate for the lower density. To generate a linear gradient, the solutions in the two chambers must have the same volume and they must be in hydrostatic equilibrium, i.e. when there is no flow out of the apparatus, there must be no tendency for fluid to flow from one compartment to the other. The density compensator insert ensures that these conditions are met.

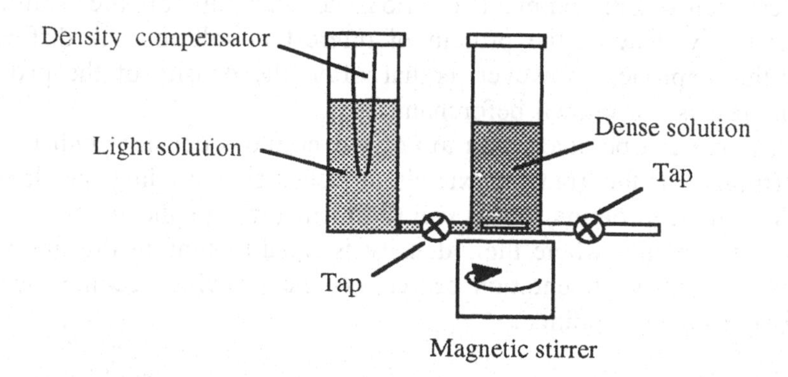


*Figure 24.* A gradient generator for use with sucrose or Ficoll.

An elegant new approach to gradient making, the tilted tube method, was introduced by Coombs and Watts[20]. In this method, the light solution is layered on top of the heavy solution directly in the centrifuge tube. The tube is then tilted and rotated, resulting in the rapid formation of a smooth gradient (*Figure 25*).

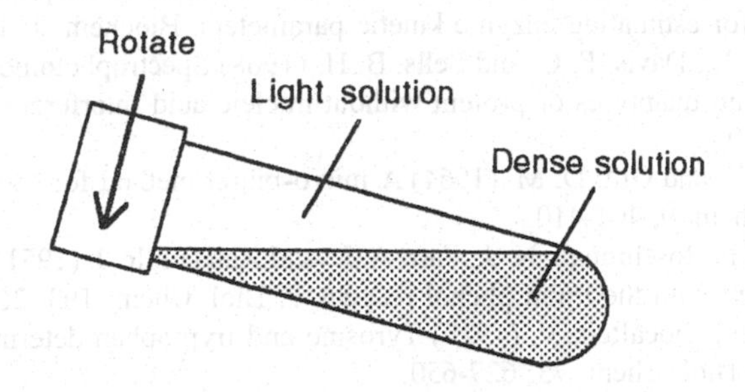


*Figure 25.* Rapid gradient generation by rotating a tilted tube.

"Nycodenz", a product of Nyegaard Diagnostica, of Norway, consists of a substituted tri-iodobenzene ring linked to a number of hydrophilic groups. It has a *MW* of 821, is non-ionic and has a density of 2,1 g/ml. With Nycodenz, or caesium chloride, a technique known as equilibrium isodensity centrifugation (isopycnic focusing) can be used. In this technique, the sample is mixed homogeneously with a concentrated solution of Nycodenz (or caesium chloride) and centrifuged. The Nycodenz (or caesium chloride) tends to sediment out slightly, generating a density gradient. At any given speed of centrifugation an equilibrium is established between sedimentation of the dense compound and back diffusion, driven by the concentration differences arising from the sedimentation. Protein molecules distribute themselves in the gradient, the density field driving them to a region where the solution density is equal to their own buoyant density. This is comparable to isoelectric focusing (IEF), and the Nycodenz plays a role analogous to that of ampholytes in IEF.

*References*

1. Dennison, C. (1988) A simple and universal method for making up buffer solutions. Biochem. Educ. 16, 210–211.
2. Ellis, K. J. and Morrison J. F. (1982) Buffers of constant ionic strength for studying pH-dependent processes. Methods Enzymol. 87, 405-426.
3. Scopes, R. K. (1994) *Protein Purification: Principles and Practice. 3rd Ed*, Springer-Verlag, New York, pp326-330.
4. Dehrmann, F. M., Coetzer, T. H. T., Pike, R. N. and Dennison, C. (1995) Mature cathepsin L is substantially active in the ionic milieu of the extracellular matrix. Arch. Biochem. Biophys. 324, 93-98.
5. Perrin, D. D. and Dempsey, B. (1974) *Buffers for pH and metal ion control.* Chapman and Hall, London.

6. Segel, I. H. (1976) in *Biochemical Calculations*, 2nd Ed, John Wiley and Sons, London.
7. Eisenthal, R. and Cornish-Bowden, A. (1974) The direct linear plot. A new graphical procedure for estimating enzyme kinetic parameters. Biochem. J. 139, 715-720.
8. Groves, W. E., Davis, F. C. and Sells, B. H. (1968) Spectrophotometric determination of microgram quantities of protein without nucleic acid interference. Anal. Biochem. 22, 195-210.
9. Itzhaki, R. F. and Gill, D. M. (1964) A micro-biuret method for estimating protein. Anal. Biochem. 9, 401-410.
10. Lowry, O. H., Rosebrough, N. J., Farr, A. L. and Randall, R. J. (1951) Protein measurement with the Folin phenol reagent. J. Biol. Chem. 193, 265-275.
11. Folin, O. and Ciocalteu, V. (1927) Tyrosine and tryptophan determination in proteins. J. Biol. Chem. 73, 627-650.
12. Peterson, G. L. (1979) Review of the Folin phenol protein quantitation method of Lowry, Rosebrough, Farr and Randall. Anal. Biochem. 100, 201-220.
13. Smith, P. K., Krohn, R. I., Hermanson, G. T., Mallia, A. K., Gartner, F. H., Provenzano, M. D., Fugimoto, E. K., Goeke, N. M., Olsen, B. J. and Klenk, D. C. (1985) Measurement of protein using bicinchoninic acid. Anal. Biochem. 150, 76-85.
14. Wiechelman, K. J., Braun, R. D. and Fitzpatrick, J. D. (1988) Investigation of the bicinchoninic acid protein assay: identification of the groups responsible for color formation. Anal. Biochem. 175, 231-237.
15. Bradford, M. M. (1976) A rapid and sensitive method for the quantitation of microgram quantities of protein utilizing the principle of protein dye-binding. Anal. Biochem. 72, 248-254.
16. Read, S. M. and Northcote, D. H. (1981) Minimization of variation in the response to different proteins of the Coomassie Blue dye-binding assay for protein. Anal. Biochem. 116, 53-64.
17. Dennison, C. Moolman, L., Pillay, C. S. and Meinesz, R. E. (2000) Use of 2-methylpropan-2-ol to inhibit proteolysis and otherv protein interactions in a tissue homogenate: an illustrative application to the extraction of cathepsins B and L from liver tissue. Anal. Biochem. 284, 157-159.
18. Storrie, B. and Madden, E. A. (1990) Isolation of subcellular organelles. Methods Enzymol. 182, 203-225.
19. Chervenka, C. H. (1969) *A manual of methods for the ultracentrifuge*. Beckman Instruments Inc., Palo Alto.
20. Coombs, D. H. and Watts, R. M. (1985) Generating sucrose gradients in three minutes by tilted tube rotation. Anal. Biochem. 148, 254-259.

## 3.7 Chapter 3 study questions

1. What is the primary measurement in the assay of an enzyme?
2. How can $V_o$ be determined from this primary measurement?
3. Left to itself, when will an enzyme catalysed reaction stop?
4. If [enzyme] is doubled, what happens to $V_o$?
5. Is $V_o$ affected by the amount of substrate present?
6. What [S] is best in an enzyme assay?
7. Describe the effect of temperature on $V_o$.
8. What is the object of extracting a protein? (Note "extraction" ≠ "isolation")
9. How may cells be lysed by osmosis?
10. Describe the physical principle common to the operation of both pestle and blender type homogenisers.
11. Why is the turbine of a water-pump windmill usually located on top of a tower?
12. A car travelling at speed over a gravel road gets dusty. Why doesn't the wind, resulting from the car's forward motion, blow the dust away?
13. What does the word "acceleration" mean?
14. Two identical particles are in a solution undergoing centrifugation, one near the top of the solution and one halfway down. i) Which one of these will reach the bottom first? ii) Will they both be completely sedimented in the time described in the equation given on p51?
15. You are following a centrifugation method, which prescribes centrifugation for 1 h at a given rpm. You wish to leave the lab in 30 min time for an important date. What can you do about this?
16. Why is it usually necessary to "wash" the pellet obtained in a fractionation by differential centrifugation? How might this "washing" be effected?
17. In real time, dairy cream has a negative sedimentation under real or artificial gravity. Explain.

# Chapter 4

## Concentration of the extract

Proteins are most efficiently extracted into dilute solution, whereas subsequent handling is more convenient if the protein is present in a relatively small volume. The first step, following the extraction of the protein into solution is, therefore, usually to concentrate it into a smaller volume. The concentration method may be non-specific, in which case only the water is removed and all non-volatile molecules are concentrated. Alternatively, it may be non-specific with respect to large molecules, i.e. the water and all small molecules are removed and all large molecules, including all the proteins, are concentrated. Finally, the concentration may be more-or-less specific, i.e. a particular protein may be concentrated in relation to the water and other molecules, including some protein molecules.

### 4.1 Freeze drying

Freeze drying is a method for the removal of water from a sample kept at low temperature, the water being removed directly from ice into the vapour phase by sublimation. It is a non-specific method as all of the non-volatile solutes are concentrated.

A major use of freeze-drying is for long term storage (preservation) of proteins or other biological samples. By reducing the water content to a very low level, microbial growth is inhibited and spoilage of the stored material is prevented. Aqueous-phase chemical reactions are also inhibited and this helps to preserve the sample. It may be noted that if the water is not removed, a temperature of *ca.* -70°C is required in order to stop aqueous-phase reactions, i.e. a deep-freeze at -20°C is not cold enough.

Freeze drying may destroy the activity of some enzymes and, if it is important to retain the activity, this should always be checked before freeze-drying is used to preserve a particular protein.

### 4.1.1 Theoretical and practical considerations in freeze-drying

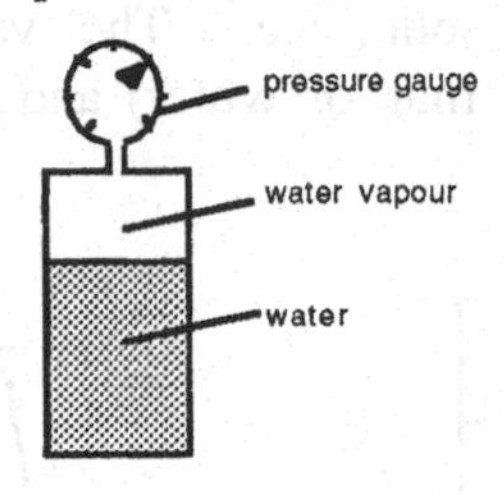


*Figure 26.* Measurement of the vapour pressure of water.

In order to understand how a freeze-dryer works, it is important to understand the concept of vapour pressure. What is "vapour pressure"? Consider the set-up shown in *Figure 26*; a sealed container containing only a liquid (water) and its vapour, i.e. with no other gases present, and with a pressure gauge to monitor the (vapour) pressure. In this way, the vapour pressure can be measured as the temperature of the set-up is changed. If this is done, it will be found that the measured pressure (the vapour pressure) changes with the temperature as shown in *Figure 27*. If there are other gases present then the vapour will contribute a part of the total pressure, i.e. the total pressure will be the vapour pressure at that temperature plus the (partial) pressures of the other gases.

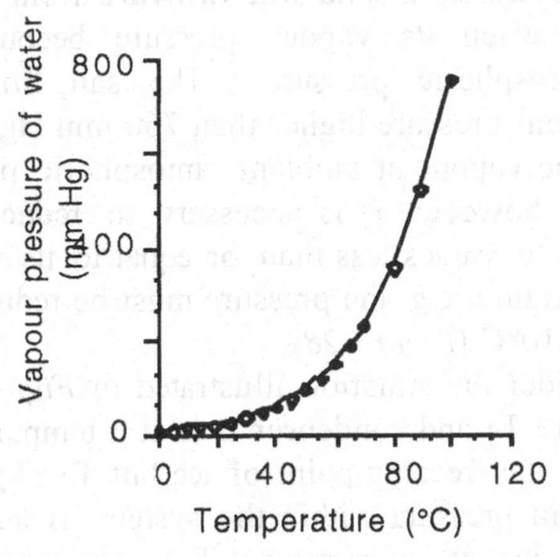


*Figure 27.* The vapour pressure of water as a function of temperature.

Notice that at 100°C, the standard boiling point of water, the vapour pressure of water is 760 mm Hg, which is the standard atmospheric pressure. This illustrates the important principle that a liquid will boil when its vapour pressure becomes equal to the environmental atmospheric pressure.

At *ca.* 0°C (at pressures of ambient or below), water undergoes a phase change from a liquid to a solid, ice. The vapour pressure of ice is relatively low (compared to that of water) and is asymptotic to zero as shown in *Figure 28*.

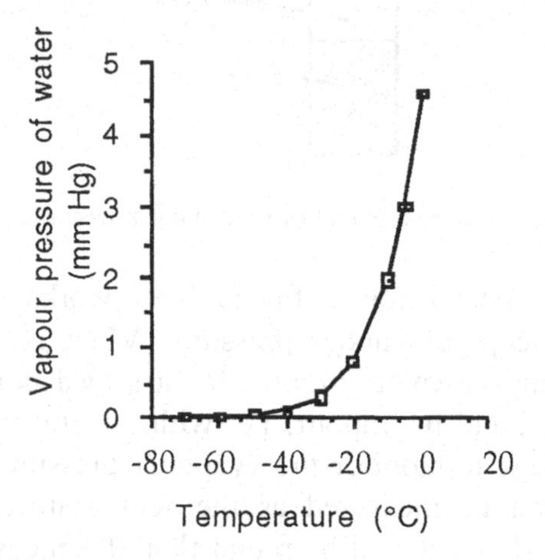


*Figure 28.* The vapour pressure of ice as a function of temperature.

Just as a liquid will boil when its vapour pressure becomes equal to the environmental pressure, so a solid will **sublime** from the solid state to the vapour state when its vapour pressure becomes equal to the environmental atmospheric pressure. The salt, *sal ammoniac*, for example, has a vapour pressure higher than 760 mm Hg and thus sublimes from the solid to the vapour at ambient atmospheric pressures. In order for ice to sublime, however, it is necessary to reduce the pressure to which it is exposed to values less than or equal to its vapour pressure at its particular temperature, e.g. the pressure must be reduced to $\leq$ 1.95 mm Hg. if the ice is at -10°C (*Figure 28*).

Therefore, consider the situation illustrated in *Figure 29*, where flask A is at a temperature $T_1$ and condenser B is at a temperature $T_2$, both $T_1$ and $T_2$ being below the freezing point of ice but $T_1 > T_2$. It follows then that if P. the overall pressure within the system, is less than $V_pT_1$ (the vapour pressure of ice at temperature $T_1$) and more than $V_pT_2$ (the vapour pressure of ice at temperature $T_2$), ice will sublime in flask A and condense in B.

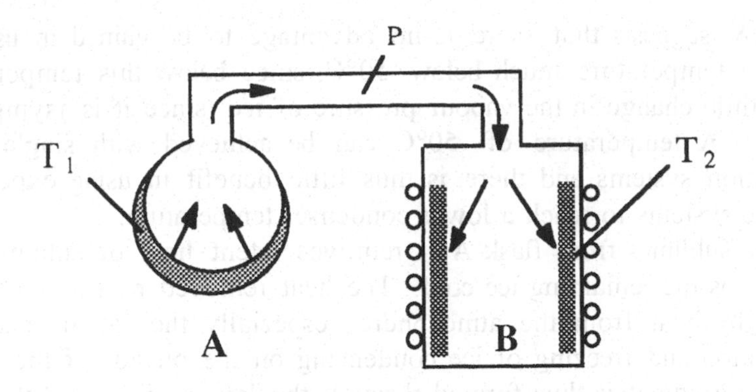


*Figure 29.* A simple freeze-dryer.

In a practical freeze-dryer, $T_1$ will be about -10°C and $T_2$ will be about -50°C. $V_pT_1$ will thus be about 1.95 mm Hg and $V_pT_2$ will be about 0.029 mm Hg (*Figure 28*). "P", the pressure measured on the vacuum gauge, must therefore be between 1.95 mm Hg (1,950 microns) and 0.029 mm Hg (29 microns): in practice it is usually between 500 and 100 microns, when the freeze dryer is operating properly.

With regard to the nett transport of water vapour from A to B an analogy can be drawn with electricity, i.e. where in electricity,

$$\text{Current} = \frac{\text{Voltage}}{\text{Resistance}},$$

In freeze-drying,

$$\frac{\text{Mass of water vapour transported}}{\text{unit time}} = \frac{\text{Vapour pressure difference}}{\text{Constant x Pressure of permanent gases}} \quad 3.1$$

The constant in the "resistance" term in equation 3.1 describes in part the geometry of the piping system connecting flask A and condenser B and is minimal when this is short and wide. To achieve a maximal rate of freeze–drying, therefore, it is necessary to establish:-

- A maximal vapour pressure difference between the sample and the condenser.
- a minimum pressure of permanent gases, and,
- a minimum value of the "piping geometry constant".

Most of these factors are fixed by the design of a particular machine but it is useful for the researcher to have an appreciation of their influence. For example, the first item, considered in conjunction with

*Figure 28*, suggests that there is no advantage to be gained in using a condenser temperature much below -50°C, since below this temperature there is little change in the vapour pressure of ice (since it is asymptotic to zero). A temperature of -50°C can be achieved with single-stage refrigeration systems and there is thus little benefit in using expensive two-stage systems to reach a lower condenser temperature.

As ice sublimes from flask A it removes latent heat of sublimation, which keeps the remaining ice cold. The heat removed by sublimation is replaced by heat from the atmosphere, especially the latent heats of condensation and freezing of ice condensing on the outside of the flask. A thermal gradient is thus formed through the layers of ice and the wall of the freeze-drying flask (*Figure 30*), and a dynamic equilibrium is established in which the rate of heat input to the system is balanced by the rate of heat loss. At equilibrium, the rate of heat input is the factor limiting the rate of freeze-drying. However, a limit to this rate of input is determined by the point at which the sample melts.

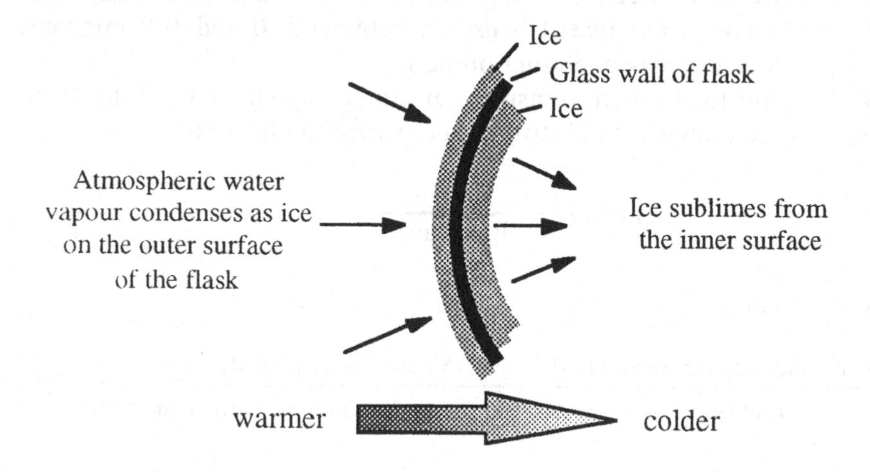


*Figure 30.* Thermal gradient across a freeze-drying flask and its associated ice layers.

The nature of the system has the following practical ramifications:-

- The greater the area over which the sample is spread, the greater will be the rate of heat input and the faster will be the rate of freeze-drying;
- The sample layer should be as thin as possible since if it is too thick there is a risk of the sample melting on its outermost surface, due to the thermal gradient. (Note: If this should happen, the sample should not be re-frozen as the melted sample, being trapped between the flask

wall and the frozen part of the sample, might crack the flask as it expands upon re-freezing.)

- The flask should be made of a material with a high thermal conductivity: thus glass is commonly used.

When the freeze-dryer is in operation, there is a flow of heat through the system, as illustrated in *Figure 31*.

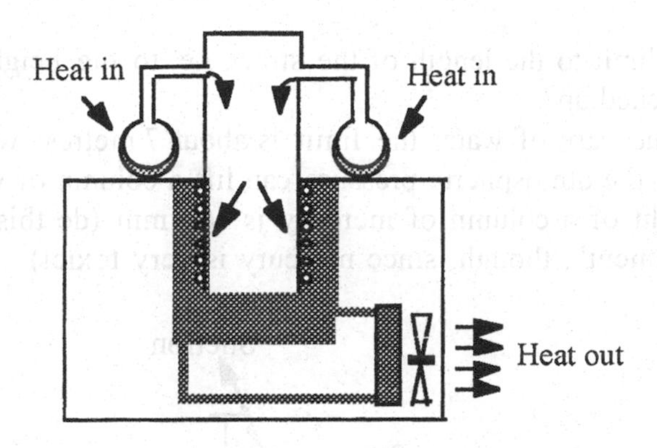


*Figure 31.* The heat flow through a freeze dryer in operation.

Atmospheric heat enters the freeze drying flask, via the ice condensing on the outside of the flask. The heat is transferred through the flask and is removed by the subliming water vapour. When this water vapour condenses, in the condenser, the heat is transferred to the refrigerant gas and is ultimately returned to the atmosphere via the radiator of the refrigeration unit.

### 4.1.2 Some tips on vacuum

Many people have confused thoughts about vacuums and so a few words on the subject may be useful. A vacuum may be defined as any pressure less than the prevailing ambient pressure. Since vacuums are defined in terms of a pressure differential, clear thinking is easier if one thinks only in terms of pressure, which has a range from zero to "infinity".

Ambient pressure is usually about one atmosphere, or 760 mm Hg, or about 15 p.s.i. ("p.s.i.", or pounds per square inch, is an old measure but one which is easy to visualise. 1 p.s.i. ≈ 6.89 kPa). An absolute vacuum, which is practically unattainable, corresponds to a pressure of zero. The theoretical maximum pressure differential across the walls of a vessel in the atmosphere but "containing" an absolute vacuum is therefore one atmosphere, i.e. 760 mm Hg or about 15 p.s.i. This is not a large differential, as pressures go, and there is clearly *no* truth in the belief that, "if one sucks hard enough, almost any vessel can be made to collapse!"

---

**How hard can one "suck" on water?**

Everyone is familiar with the process of drinking water from a container by using a straw.

**Q**: Is there a limit to the length of the straw, i.e. to the height that the water can be sucked up?

**A**: Yes. In the case of water the limit is about 7 metres, which is the height to which the atmospheric pressure can lift a column of water. The equivalent height of a column of mercury is 760 mm (do this only as a "thought experiment", though, since mercury is very toxic!)

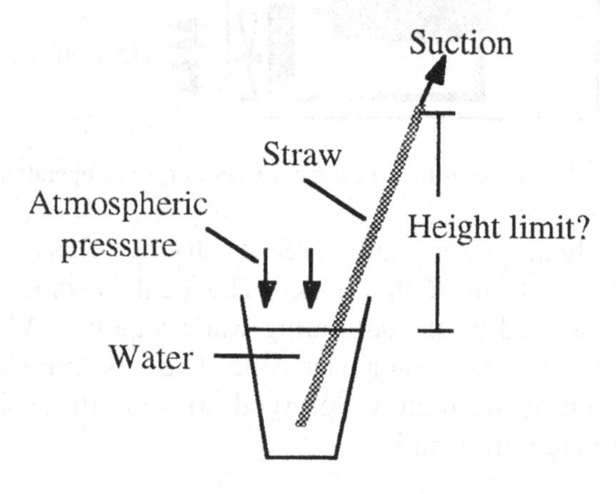


*Figure 32.* How far can water be sucked up?

This is why farmers have their water pumps near the bottom of the valley, near the river. The pump can only "suck" the water up a limited distance (about 7 metres vertical height) but it can "push" it up much further than this.

**Q**: What limits the water from being sucked up higher than 7 meters?

**A**: The water will boil (i.e. be converted to vapour) when the pressure applied to it becomes less than its vapour pressure at the prevailing temperature.

Refer back to the discussion on cavitation on p46.

---

In engineering terms the point to remember about pressure differentials is that the resultant *force* is a function of the *area* over which the differential exists. Consider, for example, the force acting upon the perspex lid of a typical laboratory freeze dryer. If the lid is 8 inches in diameter (1 inch = 2.54 cm) it will have an area of *ca.* 50 square inches. At a pressure differential of 15 p.s.i., the *force* acting on the lid is equivalent to *ca.* 750 pounds weight (*ca.* 340 kg): no wonder it bows inward. If one could get a good enough grip, one could lift the entire machine by its lid once the vacuum was established!

On the other hand, what is a high vacuum and how is it different to a "non-high vacuum"? An atmosphere corresponds to a pressure of 760 mm Hg. A pressure of 1 mm Hg is not really in the high vacuum range but the *force* applied to the system would be $^{759}/_{760}$ of the ultimate force. High vacuum pumps "split" the last mm of Hg and 0.5 mm Hg (500 microns) requires a high-vac pump, but the forces on the system will increase by only 0.5/760 or 0.07%, a negligible amount!

A generalisation can therefore be made: that if a system is structurally strong enough to withstand a "moderate" vacuum of 1 mm Hg (1000 microns) it will probably withstand any possible high vacuum! (Remember that attaining high vacuums is like splitting hairs!). In practical terms this means that one should not be too nervous of flasks imploding under vacuum. A flask is more likely to break due to thermal stress or mechanical abuse (point impacts) than under vacuum loads *per se*. Nevertheless one should be aware that the likelihood of a flask failing, from whatever cause, increases with the size of the flask.

## 4.2 Dialysis

Dialysis is the term used to describe the *diffusion* of solutes through semi-permeable membranes when the membrane forms the boundary between solutions of different concentrations. The membrane acts as an inert sieve with a certain average pore size. The pores result from the random distribution of the fibres making up the dialysis membrane.

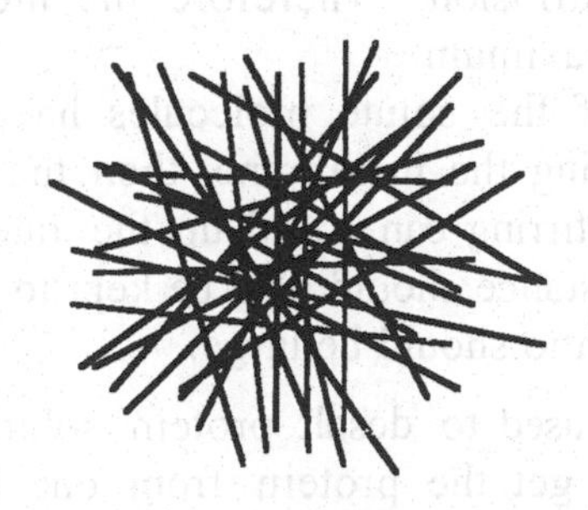

*Figure 33.* The random distribution of (cellulose) fibres in a dialysis membrane.

A "pore" corresponds to a space bounded by fibres. Clearly the pores are not all of the same size: there is a normal distribution of pore sizes. Molecules with a molecular radius larger than the largest pore size of the membrane will be completely retained while those with smaller radii will pass through more or less easily depending on their size. *Figure 33* shows a 2-dimensional representation but it must be realised that pores are 3-dimensional.

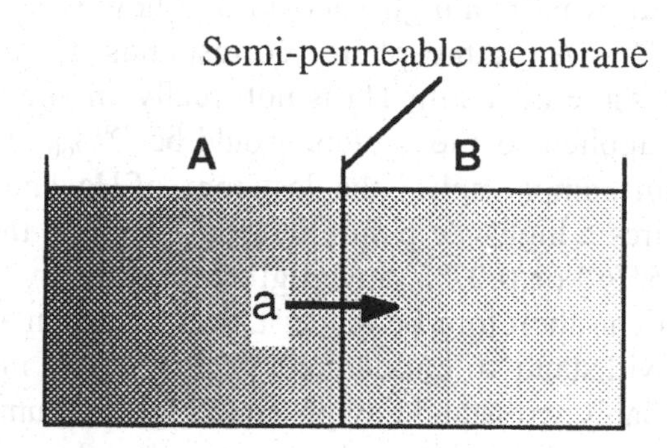


*Figure 34.* Dialysis across a semi-permeable membrane.

With reference to *Figure 34*, consider a small solute "a", initially in compartment "A" which is separated from compartment "B" by a semi-permeable membrane. As the initial concentration of "a" in A is greater than its concentration in B, "a" will diffuse from A→B.

According to Fick's law of diffusion (See p10) the rate of diffusion will be affected by the following factors:-

- *The concentration differential across the membrane.* Stirring of both solutions, if possible, and regular changing of solution B will ensure that $[a]_A >> [a]_B$ and thus the rate of diffusion will be kept at a maximum.
- *Surface area.* The larger the surface area of the membrane, the faster the overall rate of diffusion. Therefore the membrane area should always be kept at a maximum.
- *Solution volume.* If the solute molecules have to diffuse a long distance before reaching the membrane, then the rate of dialysis will be relatively slow. Stirring can speed up the rate of transfer to the membrane, but the distance should also be kept to a minimum, i.e. the surface area:volume ratio should be large.

Dialysis is typically used to desalt protein solutions, or to effect a buffer exchange, i.e. to get the protein from one buffer solution into another (Note that "desalting" and "buffer exchange" are really the same process, in the former the second "buffer" is simply distilled water).

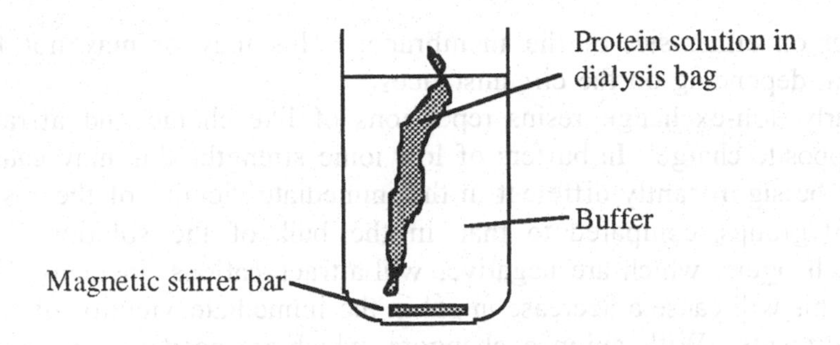


*Figure 35.* Dialysis using a visking dialysis bag.

Dialysis can be done in various ways, but in the laboratory it is most commonly done using "visking" tubing. This is a cellulosic material re-constituted into tubular form, dried, and supplied in rolls. A length can be cut from the roll, hydrated by immersion in water for several minutes, and clamped or knotted at one end to form a sealed "dialysis bag". The protein is introduced into this bag and the open end is sealed by clamping or knotting. The dialysis bag is immersed in a large volume of distilled water or buffer for several hours at 4°C to effect exchange of the permeable ions and molecules (*Figure 35*), the dialysis solution being changed at intervals (every few hours).

During dialysis, water enters the dialysis bag due to the osmotic pressure of the protein solution. For this reason a dialysis bag must not be filled, but a potential space must be left to accommodate the increasing volume of the protein solution (see Section 4.2.3). Note that if the dialysis bag is sealed with knots, the knot should be tightened by pulling only on the outside, not on the bag side of the knot, to avoid stretching the bag and thus distorting the pores.

### 4.2.1 The Donnan membrane effect

The Donnan membrane effect[1] describes the phenomenon whereby a charged macromolecule, constrained by a semi-permeable membrane, causes an asymmetrical distribution of permeable ions on either side of the membrane. The nett effect is to cause an apparent movement of ions, having the same charge as the protein, away from the protein, i.e. if the protein is positively charged, there will be a lower concentration of small cations in the compartment containing the protein than in the compartment on the other side of the membrane, and *vice versa*. In buffers, the Donnan effect is not very significant, but when a protein is dialysed against distilled water the Donnan effect can cause significant pH differences on either side of the membrane. This may or may not be significant, depending on the circumstances.

Similarly, ion-exchange resins repel ions of like charge and attract ions of opposite charge. In buffers of low ionic strength, this may cause the pH to be significantly different in the immediate vicinity of the resin substituent groups, compared to that in the bulk of the solution, i.e. cation exchangers, which are negative, will attract cations, including $H^+$ ions, and this will cause a decrease in pH in the immediate vicinity of the resin substituents. With anion exchangers, which are positive, hydroxyl ions are attracted and the pH around the substituents is consequently higher than in the bulk solution.

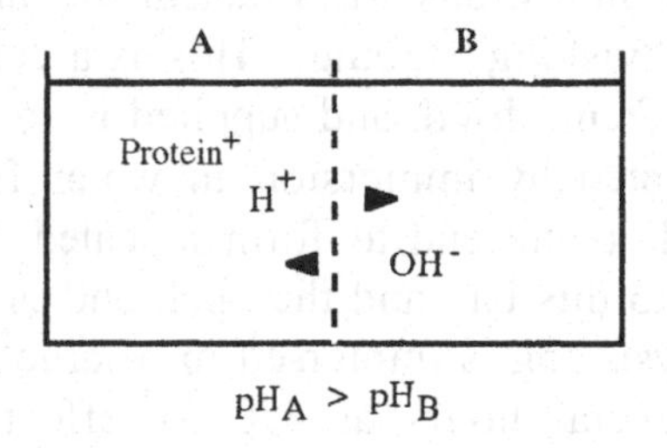


*Figure 36.* The Donnan membrane effect.

### 4.2.2 Counter-current dialysis

A very efficient form of dialysis, often used in automatic analysers, is counter-current dialysis (CCD). In this, a stream of the solution to be dialysed is arranged to flow through a thin, convoluted, channel on one side of a dialysis membrane, and the dialysing solution is arranged to flow in the opposite direction through a corresponding thin channel on the other side of the dialysis membrane. In CCD, a maximal concentration difference is thus maintained between the solution being dialysed and the dialysing solution. Since thin channels are used, the diffusion distance is small and so there is little need to stir the solutions. A stirring effect can be induced, by flowing the solutions at a high speed so that laminar flow breaks down into turbulent flow, but the period of dialysis per pass is reduced and the benefits, if any, have to be assessed for each case by empirically establishing the optimal flow rate.

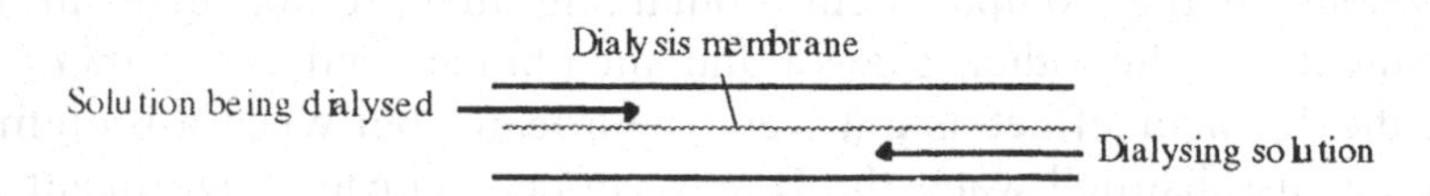


*Figure 37.* Counter current dialysis.

### 4.2.3 Concentration by dialysis (concentrative dialysis)

As mentioned above, a "complication" of dialysis is *osmosis*, which is the movement of water through a semi-permeable membrane from a solution of low osmotic pressure to a solution of high osmotic pressure. Normally the flow is into the protein solution, so that the protein solution becomes diluted during dialysis against distilled water or a buffer solution: for this reason a dialysis bag is never filled when a protein solution is dialysed. However, the flow can be reversed and the protein solution concentrated, by dialysing the protein against a solution with a higher osmotic pressure.

The dialysis bag may be simply surrounded by granular sucrose. The water flowing out of the bag will dissolve the sucrose, generating a concentrated sucrose solution with a high osmotic pressure, and this will cause further egress of water from the bag. Alternatively, the dialysis bag may be suspended in a solution of polyethylene glycol (PEG), a hydrophilic polymer.

It will be appreciated that, because water is flowing *out* of the dialysis bag in such a case, the bag can be filled completely with protein solution before concentrative dialysis. Concentrative dialysis is a specific method in the sense that only macromolecules are concentrated - all buffer salts etc. are not concentrated - but it is non-specific with respect to proteins.

An effect similar to that of concentrative dialysis can be achieved by adding a dry, reversibly hydratable gel (i.e. one that can be dried and reconstituted to have the same structure) such as Sephadex. The Sephadex xerogel will absorb water and, provided it is larger than the exclusion limit of the gel, the protein will be concentrated in the fluid between the swollen gel particles.

### 4.2.4 Perevaporation

A method of concentration using dialysis bags but which is not used much today, is perevaporation. In this method a dialysis bag containing the protein solution to be concentrated is suspended in a stream of air. Water evaporates from the outside of the bag, keeping the bag and its remaining contents cool. As the water evaporates, all of the non-volatile contents of the dialysis bag are concentrated.

An application of perevaporation which is frequently used today is in the drying of polyacrylamide gels after electrophoresis. Dried gels are mechanically strong and are more easily stored than hydrated gels. To dry the gel, a cellophane membrane is placed on either side of it and the sandwich is suspended in a stream of warm, dry, air until it is completely dry.

## 4.3 Ultrafiltration

Ultrafiltration is a technique related to dialysis, and can also be used to desalt protein solutions, effect buffer exchange, or concentrate protein solutions. It is more expensive than dialysis, however, as special equipment and membranes are required.

In this technique, pressure is applied to the solution to cause a bulk flow of water and dissolved low molecular weight solutes, through the membrane, while high molecular weight solutes are retained.

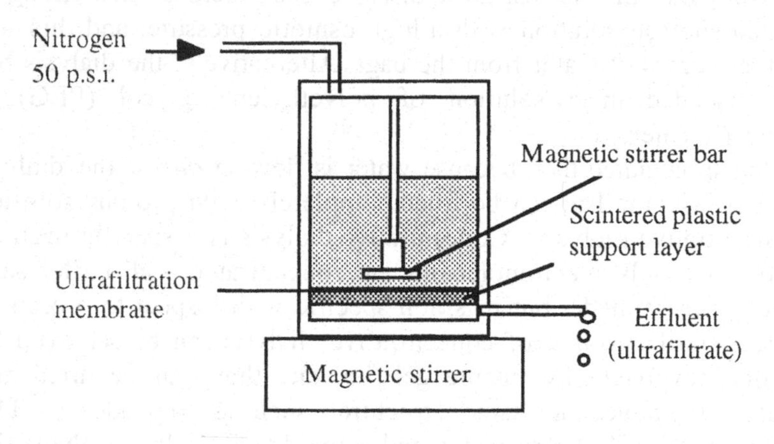


*Figure 38.* An ultrafiltration cell.

If a conventional dialysis membrane were used for ultrafiltration, it would soon become blocked with proteins trapped within the membrane. To overcome this problem, special membranes are used. These have a unimolecular sieving layer (a layer one molecule thick), supported by a much thicker support layer having a larger pore size.

Such a membrane is called an anisotropic membrane, since it is not the same in all parts. The sieving side can be distinguished from the support side because the sieving side is shiny while the support side is dull.

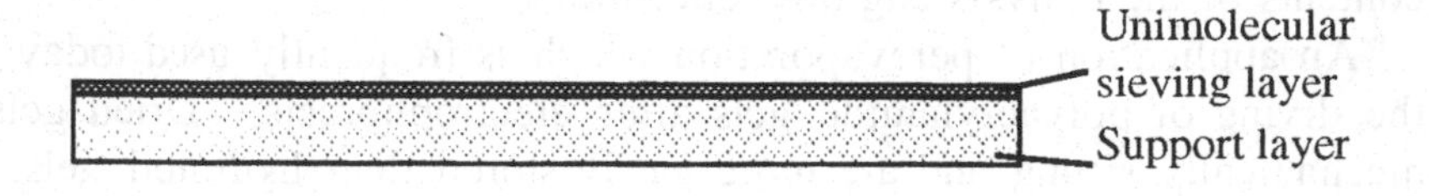


*Figure 39.* An anisotropic ultrafiltration membrane.

Whether or not a particular molecule will pass through an ultrafiltration membrane is determined at the unimolecular sieving layer. Proteins which are unable to pass through are rejected on the surface where they can easily be removed and the filter is therefore resistant to blocking.

The pressure exerted on the solution causes a flow of solvent through the membrane but immediately flow commences, a phenomenon known as *concentration polarisation* occurs. This refers to the process whereby a secondary membrane layer of retained protein is formed.

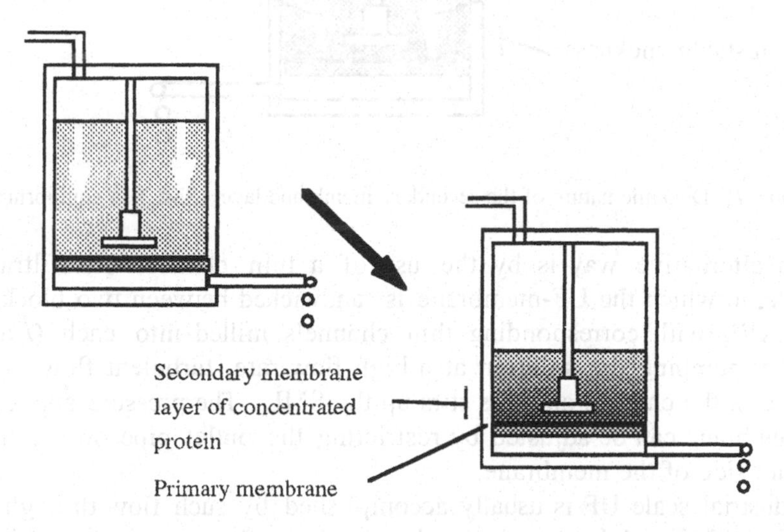


*Figure 40.* Concentration polarisation: the formation of a secondary membrane layer.

The secondary membrane layer (SML) constitutes the major resistance to flow and thus determines the flow rate. At any given pressure, an equilibrium is rapidly set up whereby the transport of macromolecules into the SML, by bulk flow of solvent, is counterbalanced by diffusion of macromolecules out of the SML. The SML thus attains a stable thickness and the flow rate remains constant.

If the applied pressure is increased, the flow rate initially increases, but this results in more macromolecules being transported into the SML. The thickness of the SML thus increases, its resistance to flow increases, and the flow rate drops, virtually to the original value. Thus the flow rate is essentially independent of the applied pressure.

Since the resistance is determined by the thickness of the SML, reducing this thickness will result in an increased flow rate. One way of decreasing the thickness of the SML is to stir the solution and thus increase the effective rate of back diffusion. This is the purpose of the stirring bar illustrated in *Figure 38*.

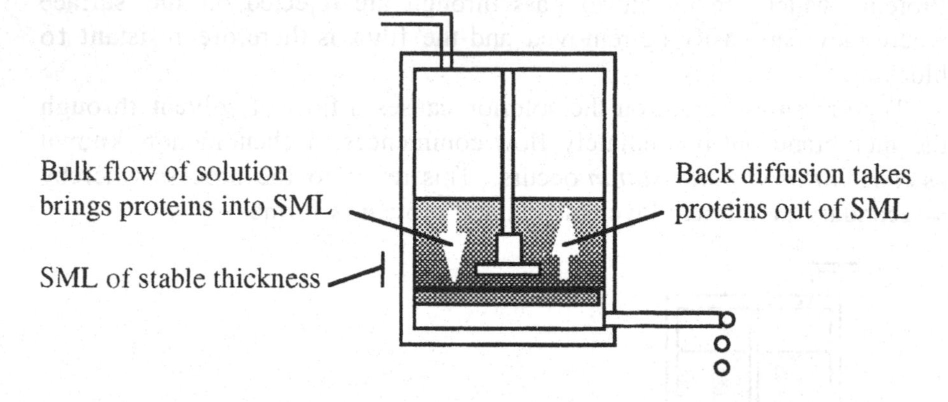


*Figure 41.* Dynamic nature of the secondary membrane layer (SML), at equilibrium.

An alternative way is by the use of a thin channel ultrafiltration module, in which the UF-membrane is sandwiched between two blocks of perspex™, with corresponding thin channels milled into each (*Figure 42*). By pumping the solution at a high flow rate, turbulent flow can be induced in the channel and this stirs up the SML. The pressure applied to the membrane can be adjusted by restricting the outlet pipe on the high-pressure side of the membrane.

Industrial scale UF is usually accomplished by such flow-through UF modules. The modules may be stacked, with the flow arranged in series or in parallel, and very high total membrane surface areas and overall flow rates can be obtained. An advantage for industrial scale applications is that such UF systems are one of the few protein fractionation methods that can run continuously, all other methods requiring batchwise operation.

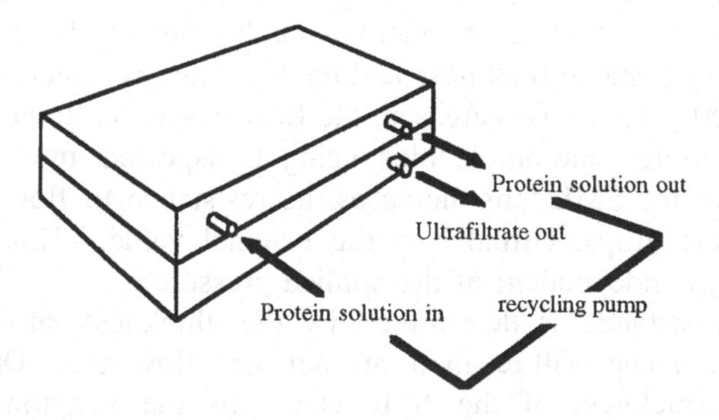


*Figure 42.* A thin-channel flow-through ultrafiltration module.

The secondary membrane determines the flow rate but the primary membrane (the anisotropic membrane) determines the *selectivity*, i.e. the size of molecules that will be retained. Primary membranes can be purchased with different "exclusion limits", ranging from 500→100,000 daltons. Conceptually, the "exclusion limit" is the molecular weight of a globular protein which will just be retained by the membrane - which is the same as the molecular weight of a protein which will just pass through the membrane. In practice, however, there is not a clear-cut distinction between the size of molecule which will be retained and that which will pass through the membrane, since the pore sizes in any membrane have a normal (Gaussian) distribution.

### 4.3.1 Desalting or buffer exchange by ultrafiltration

Ultrafiltration may be used to change the buffer in which a protein is suspended, either for another buffer or for distilled water. In either case the approach used is the same. The solution is reduced to a small volume, rediluted in the new buffer (or distilled water), reduced to a small volume again etc., the process being repeated several times, until the protein is in the new buffer only. The process is analogous to the washing of the retentate on a paper filter and the same equation applies, i.e.:-

$$x_n = x_o \left[\frac{u}{u+v}\right]^n$$

Where $n$ = number of "washings"
$x_o$ = conc. of original salts before desalting
$x_n$ = conc. of original salts after "n" washings
$u$ = volume to which sample is reduced
$v$ = volume of new solution added for each washing

Consequently, to achieve buffer exchange (or desalting) in the minimum time, the factors "u" and "v" should be kept to a minimum, since the time taken depends upon the total volume of washing solution used. (This is a useful equation to remember when rinsing one's laundry).

### 4.3.2 Size fractionation by ultrafiltration

Proteins may be fractionated into size groups by ultrafiltration, by passing the solution, successively, through membranes of decreasing pore size. The largest proteins will be retained by the membrane with the largest pore size, etc.

## 4.4 Concentration/fractionation by salting out

Salting out using ammonium sulfate is one of the classical methods in protein biochemistry. Formerly it was widely used for the fractionation of proteins, but it is not a highly discriminating method and it is unusual to get a pure fraction, using this method. Today it is rather used as an inexpensive way of concentrating a protein extract, while leaving non-protein material in solution, and any purification with respect to protein is generally regarded as a bonus.

### 4.4.1 Why ammonium sulfate?

Empirically, it has been found that polyvalent anions are more effective at salting out than univalent anions, while polyvalent cations tend to negate the effect of polyvalent anions. The best combination is therefore a polyvalent anion with a univalent cation. Anions can be arranged in a so-called "Hofmeister series", which describes their relative effectiveness in salting out at equivalent molar concentrations[2]. In decreasing order of effectiveness, the series is: citrate > sulfate > phosphate > chloride > nitrate > thiocyanate. This series also describes a decreasing tendency for the anions to stabilise protein structure. Citrate and sulfate are thus "kosmotropes", which tend to stabilise protein structure, while thiocyanate and nitrate are "chaotropes" which tend to destabilise protein structure. An ideal salt would, therefore, be citrate or sulfate combined with a univalent cation. Ammonium sulfate is most popular because it meets these criteria, is available in a pure form at low cost and is highly soluble, so that high solution concentrations can be attained.

The sulfate ion has been viewed in a number of ways, regarding how it salts out proteins, including, ionic strength effects, kosmotropy, exclusion-crowding, dehydration, and binding to cationic sites, especially when the protein has a nett positive charge (denoted $Z_H$+)[3]. All of these may play a role, depending upon the salt concentration and the pH-dependent charge on the protein.

*Ionic strength effects.* It will be noticed that the Hofmeister series goes from multivalent to univalent ions. This largely reflects the fact that the Hofmeister series is based on molarity, while ionic strength is a factor in salting out. The valency of the ion has an effect on ionic strength as can be illustrated by comparing NaCl with $(NH_4)_2SO_4$.

Ionic strength is defined as:-

$$I = \frac{1}{2}\Sigma c_i(z_i)^2$$

Where, $c_i$ = concentration of each type of ion (moles/litre)
$z_i$ = charge of each type of ion.

Thus in the case of 1 M NaCl,

$$I = \frac{1}{2}[(1 \times 1^2) + (1 \times 1^2)]$$
$$= \frac{1}{2}(1 + 1)$$
$$= 1$$

and for 1 M $(NH_4)_2SO_4$,

$$I = \frac{1}{2}[(2 \times 1^2) + (1 \times 2^2)]$$
$$= \frac{1}{2}(2 + 4)$$
$$= 3$$

Ionic strength effects come into play at low salt concentrations (0→0.2 M) and, as the name implies, they are not specific to ammonium sulfate. At low ionic strength, protein solubility is at its minimum at the protein's pI (*Figure 43*). At this pH, intramolecular electrostatic forces between oppositely charged side chains are at a maximum, protein conformation is maximally tightened and protein hydration is least. On either side of the pI, titration of ionisable groups leads to a lessening of intramolecular ionic interactions. In consequence, protein structure becomes more relaxed and hydration and solubility are increased[3].

Addition of low concentrations of salt causes a similar weakening of intramolecular ionic bonds, with similar consequences of more relaxed protein structure and greater solubility. As shown in *Figure 43*, addition of salt and altering of the pH, away from the pI, have similar, and additive, effects. The increase in solubility of protein upon addition of modest amounts of salt is known as "salting in".

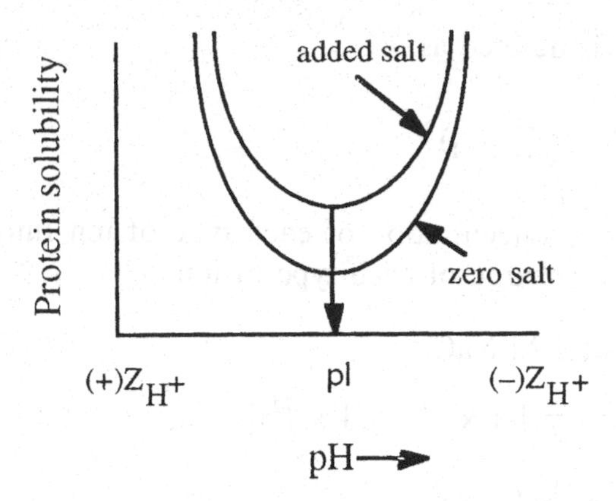


*Figure 43.* "Salting in" of proteins - the interaction of pH and ionic strength (adapted from Dennison and Lovrien - ref. 3).

*Kosmotropy.* At concentrations above 0.2 M the sulfate ion acts as a Hofmeister kosmotrope, i.e. it stabilises protein structure, and concomitantly reduces its solubility. The effect of a kosmotrope, in stabilising protein structure, can be described by the reaction:-

| Relaxed, open protein structure | → | compact, tight structure |
|---|---|---|
| (more soluble, less stable) | | (less soluble, more stable) |

Kosmotropes may be described as "pushing" if they act on the left of this reaction and "pulling" if they act on the right, in either case driving the reaction to the right.

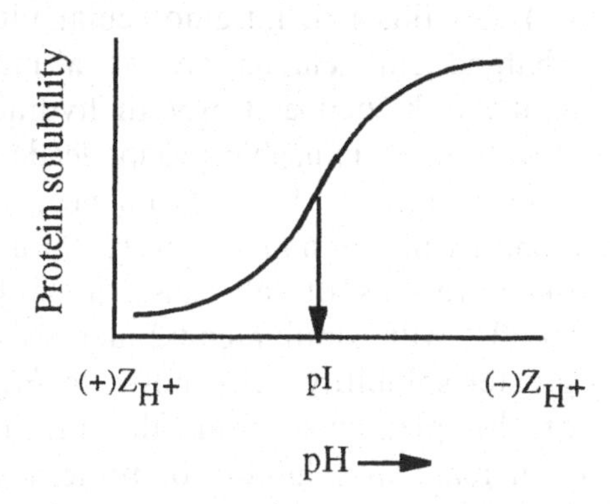


*Figure 44.* The effect of pH on the salting out of a protein by ammonium sulfate.

Sulfate can act as a pulling kosmotrope by virtue of its interaction with protein cationic sites. Consistent with this, the precipitation of proteins is usually promoted at pH values below the pI (*Figure 44*), where the protein has a maximal number of cationic sites[3]. Reinforcing its pulling effect is the fact that the sulfate ion is divalent, and so can bind to more than one cationic site at a time, and that it has a tetrahedral structure, with four oxygen atoms that can hydrogen bond to multiple sites on the protein.

Sulfate also acts as a pushing kosmotrope by virtue of its extraordinary hydration. By virtue of its hydration, the sulfate ion can act as a dehydrating agent and, in its hydrated form, as an exclusion-crowding agent. The sulfate anion has 13 or 14 water molecules in its first hydration layer and possibly more in a second layer[4]. Consequently, in salting out at, say, 3 M ammonium sulfate, the sulfate anion will have accreted to itself 40 to 45 M out of the total of 55 M $H_2O$ in neat water. In salting out, therefore, a large proportion of the water will be involved in hydrating the sulfate ions and increasing their effective radius. The large, hydrated, $[SO_4.(H_2O)_n]^{2-}$ ions crowd and exclude the proteins, pushing them into tighter, more ordered (less soluble) conformations, with lower entropy[5]. The preferential accretion of the water molecules to the sulfate ions excludes the proteins from a proportion of the water (the proportion increasing with the salt concentration), ultimately bringing them to their solubility limit.

No other salt has the combination of properties which make ammonium sulfate so effective at salting out. Consequently, when the word "salt" is used in the context of salting out, it invariably means ammonium sulfate. Similarly, the term "ionic strength" is often used loosely, when what is really meant is "the concentration of ammonium sulfate".

### 4.4.2 Empirical observations on protein salting out.

Starting from zero, increases in salt concentration initially increase the solubility of the proteins, due to salting in. With further increases in ammonium sulfate concentration, the protein solubility passes through a maximum and then decreases (*Figure 45*).

The salting out relationship is described by an empirical equation[6]:-

$$\text{Log } S = \beta - K_s I$$

where,
$S$ = protein solubility (g/l)
$I$ = ionic strength
$\beta$ and $K_s$ are constants.

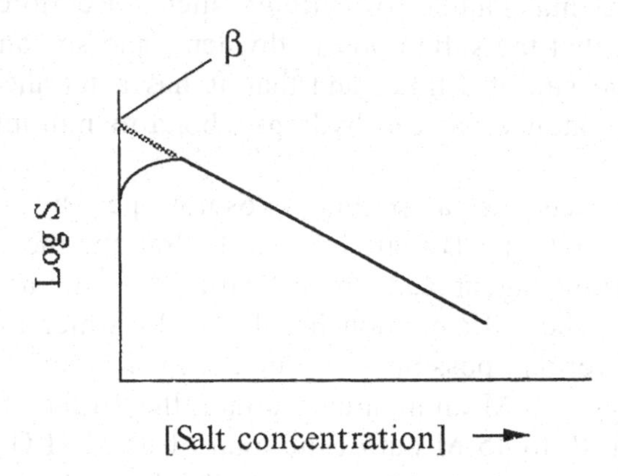


*Figure 45.* Solubility of a typical protein vs concentration of ammonium sulfate.

$K_s$, the so-called "salting out constant" (the slope of the plot in *Figure 45*), is essentially independent of temperature and pH but varies slightly with the nature of the protein. $\beta$ is markedly dependent upon the pH and temperature (*Figure 46*) and also varies markedly with the nature of the protein.

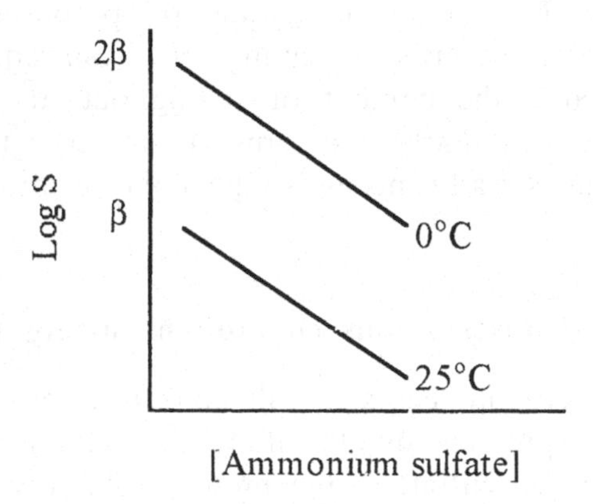


*Figure 46.* The effect of temperature on the salting out of carboxyhaemoglobin (adapted from ref. 7).

Note that a rise in temperature causes a *decrease* in $\beta$. Note also that, since $\beta$ is in units of log $S$, a unit change in $\beta$ represents a ten-fold change in solubility. Therefore, a protein will be markedly less soluble at higher temperatures and in practice it is better to conduct salting out at, say, 25°C rather than at 4°C. The sulfate ion is kosmotropic, so proteins are stabilised by the presence of $(NH_4)_2SO_4$ and a high salt concentration also inhibits microbial growth. For these reasons, also, it is less necessary than usual to work at a low temperature.

The initial concentration of a protein in solution has a major influence on the amount of $(NH_4)_2SO_4$ required to precipitate it. Proteins appear to fall into two categories, denoted type I and type II[2], depending upon how their concentration affects their salting out behaviour. For type I proteins, each protein has a characteristic precipitation curve (e.g. *Figure 47*).

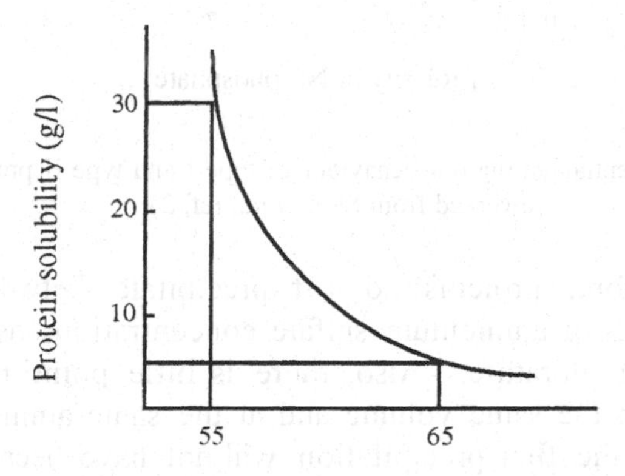


*Figure 47.* The salting out curve of carboxymyoglobin (adapted from ref. 7).

The lower the concentration of protein in solution, the more salt is required to precipitate it. In the example given in *Figure 47*, if carboxymyoglobin is present at an initial concentration of 30 g/l, it will begin to precipitate at about 55% saturation with $(NH_4)_2SO_4$, whereas at an initial concentration of 4 g/l, 65% saturation is required to begin precipitation.

Not all proteins behave in this simple way. Type II proteins[2], such as BSA and α-chymotrypsin, precipitate to an extent dependent upon their initial concentration, i.e. such proteins manifest a family of precipitation curves, each curve arising from a particular initial protein concentration (*Figure 48*).

Type I proteins have a single precipitation curve, regardless of the initial protein concentration. Type II proteins precipitate in a manner dependent upon their initial concentration.

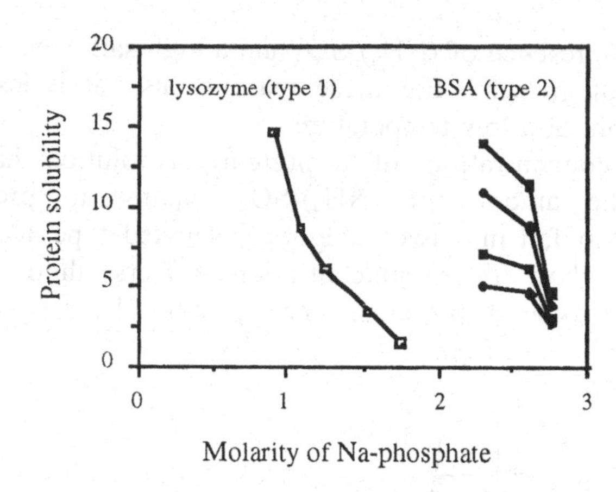


*Figure 48.* Differential salting out behaviour of type I and type II proteins. (Data reworked from Shih *et al.*, ref. 2).

Clearly, therefore, proteins do **not** precipitate between fixed and characteristic limits of ammonium sulfate concentration, as is implied in much of the older literature. Also, there is little point in repeating a precipitation, from the same volume and at the same ammonium sulfate saturation. Since the first precipitation will not have been quantitative, the concentration will be less if the protein is reconstituted in the same volume. To repeat the precipitation, the protein concentration should be readjusted to the same value as previously, which is not always practicable. In general, it is not worth repeating the precipitation as the cost, in terms of protein lost, is not justified by the increase in protein purity obtained.

Proteins may be purified from a mixture by differential precipitation at different saturations of ammonium sulfate (e.g. *Figure 49*). The protein solubility curve (*Figure 49*) has two steps, due to the precipitation of the serum albumin, followed by the carboxymyoglobin. Such perfect separation is rare, however, and it is more usual to obtain mixed fractions with, possibly, only slight enrichment of a desired protein. By altering the protein concentration it is sometimes possible to improve the separation, e.g. by diluting the solution, the points of precipitation (the peaks in the first derivative curve) will be moved to the right, to higher saturation levels. Due to differences in $K_S$, however, the peaks due to different proteins might move to different extents and the separation will thus be improved.

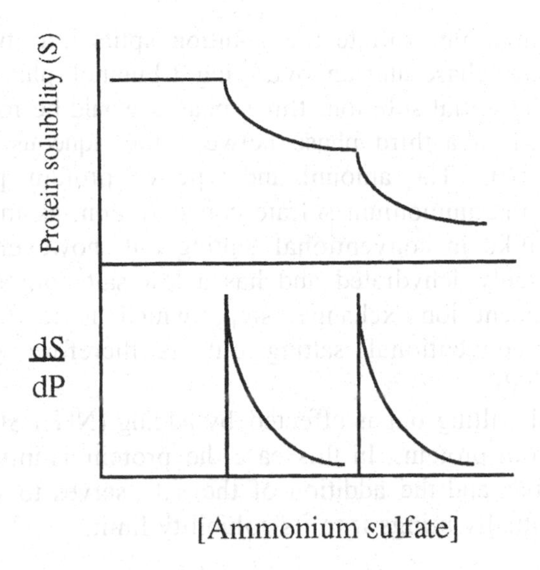


*Figure 49.* Separation of human serum albumin and carboxymyoglobin by salting out (from Dixon and Webb - ref. 7).

To summarise, in salting out the following can be manipulated;-

- *pH.* It is best to use a pH below the pI of the desired protein where precipitation is maximal.
- *Temperature.* Theoretically, the best temperature is the highest temperature at which the protein is stable. β decreases as the temperature is increased and there is therefore greater precipitation at higher temperatures. For practical purposes, room temperature (25°C) is adequate.
- *Protein concentration.* The difficulty here is that the direction of any effects cannot be predicted in advance.

The resolving power of salting out is not high and so it is now commonly used mainly as a means of concentrating proteins from dilute extracts, while leaving non-protein molecules in solution. It is also generally used early in an isolation, immediately after preparation of the extract.

### 4.4.3 Three-phase partitioning (TPP)

Three-phase partitioning (TPP) is a method in which proteins are salted out from a solution containing a mixture of water and t-butanol[3,8,9]. t-Butanol is infinitely miscible with water but upon addition of sufficient ammonium sulfate the solution splits into two phases, an underlying aqueous phase and an overlying t-butanol phase. If protein was present in the initial solution, three phases would be formed, protein being precipitated in a third phase between the aqueous and t-butanol phases (*Figure 50*). The amount and type of protein precipitated is dependent upon the ammonium sulfate concentration, as in conventional salting out. Unlike in conventional salting out, however, the protein precipitate is largely dehydrated and has a low salt content. Desalting before a subsequent ion-exchange step, which is a time-consuming necessity after conventional salting out, is therefore generally not necessary with TPP.

Conventional salting out is effected by adding $(NH_4)_2SO_4$ to a purely aqueous solution of protein. In this case the protein is initially hydrated and is thus soluble, and the addition of the salt, serves to dehydrate the protein and eventually brings it to its solubility limit.

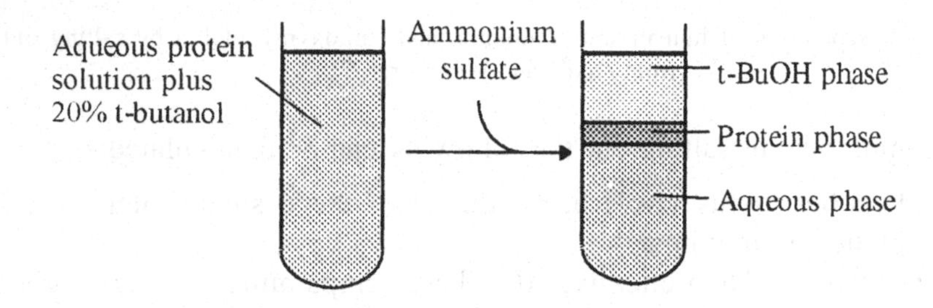


*Figure 50.* Three-phase partitioning.

In TPP, t-butanol may be first added to the aqueous solution of protein to about 20-30%. It is believed that this results in the protein equilibrating with the solvent (water) and the co-solvent (t-butanol). The protein thus becomes partially hydrated and partially t-butanolated, in proportion to the relative abundance of the solvents in the mixture.

Upon addition of $(NH_4)_2SO_4$, water is abstracted by the salt ions (especially the $SO_4^{2-}$ ions) as these become hydrated. The salt apparently has a higher affinity for water than for t-butanol, and thus preferentially sequesters the water. In the absence of protein, this results in the solution dividing into two phases, as some of the water is made "unavailable" to the t-butanol. If protein is present, the protein equilibrates with the new proportions of solvent and co-solvent available to it. Upon addition of further $(NH_4)_2SO_4$, eventually the amount of water available to the protein becomes insufficient to keep the protein in solution, and it precipitates. At this point, however, the protein will be largely t-butanolated. This results in the protein having a reduced density and so, when it precipitates out of solution, it will float on top of the aqueous layer. This contrasts with the situation in conventional salting out where the dehydrated protein normally sinks, indicating that it is *more* dense than the solution. In conventional salting out, as the concentration of $(NH_4)_2SO_4$ increases, the solution density increases and the difference in density between the solution and the precipitate decreases, until the point is reached where it is no longer possible to sediment the precipitate. In the case of TPP, however, the precipitate is *less* dense than the solution and so, with increasing $(NH_4)_2SO_4$ concentration, the precipitate floats more and more easily.

In non-aqueous environments, α-helices are favoured and so, during TPP, as the proportion of t-butanol increases, the protein conformation may become distorted as it acquires a greater proportion of α-helices. This distortion leads to the denaturation of many proteins, which may be a disadvantage. On the other hand, if the protein of interest is able to survive TPP, then it is likely that TPP will effect a purification by the denaturation of impurities, as well as by its fractionating ability. Selective denaturation of contaminating protein has been put to good effect in the isolation of cathepsin D[10] and a number of erythrocyte proteins[11], where the major problem was the presence of a large excess of haemoglobin.

To effect a fractionation of a protein mixture by TPP, about 20-30% t-butanol can be added to the protein mixture in aqueous solution and increments of $(NH_4)_2SO_4$ are added, the interfacial precipitate being concentrated by centrifugation and removed for analysis after each increment of $(NH_4)_2SO_4$.

Empirically, it has been found that, in TPP;-

- Proteins precipitate in order of their molecular weight, i.e. larger proteins precipitate before smaller proteins, present at the same concentration (probably due to molecular crowding by hydrated sulfate ions).
- Proteins are most readily precipitated into the third phase when they have a positive charge (as positively charged proteins will bind more negative sulfate ions).
- Proteins are most soluble, after TPP (i.e. denaturation is minimal), when TPP is done at the pI of the protein.
- The protein concentration has a marked influence: generally the greater the concentration, the more easily the protein will precipitate.
- Temperature has little effect, in the range 0°C to 25°C.

It should be noted that t-butanol is unusual in that it is an organic solvent which tends not to denature proteins. TPP can therefore be done at room temperature, which is fortunate because t-butanol solidifies at about 25°C, and is most conveniently used above this temperature.

#### 4.4.3.1 Homogenisation in t-butanol

A problem with tissue homogenisation is the possible generation of artefacts, either by proteolysis or by other protein/protein interactions. As stated above, t-butanol is an unusual organic solvent in that it does not denature proteins, but at about 30% (v/v), t-butanol inhibits the activity of enzymes[12] (at least of all the enzymes thus far tested). In most cases this inhibition is reversible; upon removal of the t-butanol, either by dialysis or phase separation by TPP, the enzyme activity is restored. Thus, when homogenisation is effected in the presence of 30% t-butanol proteases and protein/protein interactions are inhibited. The general result is fewer artefacts and a greater yield. Moreover, such a homogenate is poised for fractionation/concentration, by TPP, simply by adding increments of ammonium sulfate. Homogenisation in 30% t-butanol should always be explored in the development of a novel protein isolation.

## 4.5 Fractional precipitation with polyethylene glycol

Polyethylene glycol (PEG) (I) is a hydrophilic polymer. It is commercially available as preparations of different average molecular weights, the 6,000 and 20,000 materials being most often used for protein isolation.

$$\text{R-}(\underset{|}{\overset{\text{OH}}{\text{CH}}}\text{ - }\overset{\text{OH}}{\underset{|}{\text{CH}}})_n\text{ - R}$$

**I**

It is thought that PEG precipitates proteins by virtue of excluding the proteins from a relatively large volume of water per PEG molecule. With an increase in PEG concentration the protein is thus eventually brought to its solubility limit. It has been empirically observed that, at equimolar concentrations, large proteins are precipitated at lower PEG concentrations than small proteins. For this reason PEG precipitation is especially suited to the isolation of large proteins and even particles, such as viruses. PEG-precipitation may thus be thought of as "molecular exclusion precipitation"[13].

The effect of protein concentration has not been explored in detail but it would appear from limited data that in PEG precipitation proteins behave much as type I proteins do in salting out, i.e. the higher the protein concentration, the less precipitant is required to initiate precipitation.

After precipitation, it is difficult to separate the PEG from the protein. PEG is acetone soluble, however, and acetone precipitation of the protein can be used to separate it from the PEG. It may not always be necessary to remove the PEG since many of the subsequent uses of the protein may be tolerant of the presence of PEG. It should be noted, though, that PEG can alter the behaviour of proteins during molecular exclusion chromatography[14].

A practical difficulty in the use of PEG is that it absorbs UV light at 280 nm, and also interferes with the Lowry protein assay. Protein can, however, be measured in the presence of PEG using the biuret assay.

## 4.6 Precipitation with organic solvents

Proteins are maintained in solution by the interaction of surface hydrophilic groups with the water solvent. Consequently, if the polarity of the solvent is reduced by the addition of an organic solvent less polar than water, the protein will tend to become less soluble. Concurrently, the protein usually becomes less stable because a major contributor to protein stability is the distribution of hydrophilic and hydrophobic groups into the lowest energy conformation, based on the solvent being water. To minimise denaturation of the protein in the less polar solvent, it is usually necessary to conduct the fractionation at a low temperature.

Two solvents that have been commonly used in the fractional precipitation of proteins are ethanol and acetone. Ethanol is widely used in the low temperature fractionation of blood proteins[15], for example, while acetone was formerly commonly used to make "acetone powders", a means of preserving proteins.

To effect the precipitation of a protein with an organic solvent, the protein solution should be chilled close to 0°C and the solvent to at least -20°C. The solvent is slowly but smoothly added to the protein solution, with constant stirring to avoid the formation of high local concentrations of solvent. Occasionally, addition of the solvent leads to the formation of a milky colloidal suspension, rather than a precipitate. If this happens, addition of a drop of NaCl solution may be necessary to induce flocculation.

For the preparation of an "acetone powder", the protein precipitated with acetone is harvested by centrifugation and spread out to air dry. The latent heat of vaporisation of the acetone keeps the protein cool as it dries to a powder. Alternatively, the protein precipitate may be stored at low temperature and subsequently reconstituted by dissolving in chilled buffer solution. Some denaturation of the protein during organic solvent precipitation is unavoidable.

## 4.7 Dye precipitation

For the reaction:-

Protein in solution → Protein precipitate

most of the methods discussed above may be called "pushing" methods, in that the properties of **the solution** are changed, making these unsuitable for protein dissolution, i.e. the protein is "pushed" out of solution. If the protein is present at a very low concentration, in a large volume of solution, "pushing" methods can be quite uneconomical as the amount of precipitant required is proportional to the total solution volume, and often inversely proportional to the protein concentration. The alternative is a "pulling" method, in which properties of **the protein** are changed so that it comes out of solution. The "push/pull" terminology is due to Dr Rex Lovrien, of the University of Minnesota.

An example of a pulling method is dye precipitation or, as it has been called, "matrix co-precipitation"[16]. Proteins are kept in solution by the disposition of charged, hydrophilic, groups on their surfaces. At low pH these are mainly positive and at high pH mainly negative. Dyes, on the other hand, are typically salts of strong acids or bases with attached aromatic groups having extended conjugation, which gives rise to their colour. If a dye having a negatively charged sulfonic acid group is added to a positively charged protein, ionic bonds will form between the dye and the protein. As a result, bulky, hydrophobic groups will become attached to the protein at its previously positive sites and the protein will be precipitated out of solution. An advantage of this method is that the amount of dye required is proportional to the moles of protein present, not to the volume of solution, so it is particularly suitable for harvesting proteins from dilute solutions.

After precipitation, it is necessary to separate the protein from the complexed dyestuff. This can be accomplished either by ion-exchange or by TPP. In TPP, the high salt concentration breaks the ionic bonds and the released dye is extracted into the t-butanol layer. Since dye precipitation is a "pulling" method and TPP is a "pushing" method, the sequential application of these two could be described as a "pull-push" method.

An example of dye precipitation and a discussion of the mechanism is provided by Wu *et al.*[17]

*References*


1. Dawes, E. A. (1972) in *Quantitative Problems in Biochemistry, 5th Ed.* Longmans, Edinburgh, p94.
2. Shih, Y-C., Prausnitz, J. M. and Blanch, H. W. (1992) Some characteristics of protein precipitation by salts. Biotech. Bioeng. 40, 1155-1164.
3. Dennison, C. and Lovrien, R. (1997) Three phase partitioning: concentration and purification of proteins. Protein Expression and Purification 11, 149-161.
4. Cannon, W. R., Pettit, B. M. and McCammon, J. A. (1994) Sulfate anion in water: model structural, thermodynamic and dynamic properties. J. Phys. Chem. 98, 6225-6230.
5. Herzfeld, J. (1996) Entropically driven order in crowded solutions: Liquid crystals to cell biology. Accounts Chem. Res. 29, 31-37.
6. Cohn, E. .J. and Edsall, J. T. (1943) in *Proteins, amino acids and peptides.* Reinhold, New York.
7. Dixon, M. and Webb, E. C. (1961) Enzyme fractionation by salting out: a theoretical note. Adv. Prot. Chem. 16, 197-219.
8. Odegaard, B. H., Anderson, P. C. and Lovrien, R. E. (1984) Resolution of the multienzyme cellulase complex of *T. reesei* QM9414. J. Appl. Biochem. 6, 156-183.
9. Pike, R. N. and Dennison, C. (1989) Protein fractionation by three-phase partitioning (TPP) in aqueous/t-butanol mixtures. Biotech. Bioeng. 33, 221-228.
10. Jacobs, G. R., Pike, R. N. and Dennison, C. (1989) Isolation of cathepsin D using three-phase partitioning in t-butanol/water/ammonium sulfate. Anal. Biochem. 180, 169-171.
11. Pol, M. C., Deutsch, H. F. and Visser, L. (1989) Purification of soluble enzymes from erythrocyte hemolysates by three phase partitioning. Int. J. Biochem. 22, 179-185.
12. Dennison, C. Moolman, L., Pillay, C. S. and Meinesz, R. E. (2000) Use of 2-methylpropan-2-ol to inhibit proteolysis and otherv protein interactions in a tissue homogenate: an illustrative application to the extraction of cathepsins B and L from liver tissue. Anal. Biochem. 284, 157-159.
13. Polson, A. (1977) A theory for the displacement of proteins and viruses with polyethylene glycol. Prep. Biochem. 7, 129-154.
14. Arakawa, T. (1985) The mechanism of increased elution volume of proteins by polyethylene glycol. Anal. Biochem. 144, 267-268.
15. Curling, J. M. (1980) in *Methods of Plasma Protein Fractionation*, Academic Press, London.
16. Conroy, M. J. and Lovrien, R. E. (1992) Matrix coprecipitating and cocrystallizing ligands (MCC ligands) for bioseparations. J. Crystal Growth 122, 213-222.
17. Wu, C. W., Lovrien, R. and Matulis, D. (1998) Lectin coprecipitative isolation from crudes by Little Rock orange ligand. Anal. Biochem. 257, 33-39.

## 4.8 Chapter 4 study questions

1. What are the two major uses of freeze-drying?
2. What is meant by the "vapour pressure" of water?
3. Should a freeze drier have: a) short, wide-bore tubing, or , b) long, thin-bore tubing between the sample and the condenser? Explain.
4. A freeze-drier condenser usually operates at about -50°C. Is there any benefit in using a lower temperature? Explain.
5. What is a "vacuum"? and an "absolute vacuum"?
6. Imagine this situation. A freeze-drier is placed on top of a building 4 storeys high and switched on. A glass tube, connected to one of its ports reaches down into a beaker of water on the ground floor. Explain what you think would happen when the tap on that particular port is opened, so that the glass tube becomes evacuated.
7. Define "dialysis" and describe three factors that affect its rate.
8. Explain how dialysis can be used to concentrate a protein solution.
9. In what way does an ultrafiltration membrane differ from a dialysis membrane?
10. What is meant by "concentration polarisation"?
11. How is the flow rate of ultrafiltration affected by the applied pressure?
12. Why is ammonium sulfate a popular choice for salting out?
13. Describe the effects of, a) temperature, b) pH, and, c) protein concentration on the salting out of a protein.
14. Why do proteins float on the aqueous phase in TPP, yet sink in conventional salting out?
15. Describe one advantage of TPP over conventional salting out.

Chapter 5

# Chromatography

Several of the methods discussed in the previous chapter - ultrafiltration, salting out and TPP - besides being concentrating methods, can also be used for preparative fractionation. Similarly, ion-exchange chromatography can be used to concentrate a dilute solution. The division of the chapters between concentrating and fractionating methods is therefore somewhat arbitrary, but is based on whether a given method is more effective in concentrating or in fractionating.

The essence of preparative fractionations, as distinct from the analytical fractionations to be discussed in the following chapters, is that they are non-destructive and the product is an active protein. Also, preparative fractionations are usually done on a larger scale than analytical fractionations, but the scale is very dependent upon the particular problem being addressed.

After concentration of the extract by one of the methods discussed in Chapter 4, it is assayed for activity and analysed, for example by polyacrylamide gel electrophoresis (see Chapter 6). As mentioned in Chapter 2, in the absence of any other information regarding the protein, experience suggests that ion-exchange chromatography is the best method to use for the first preparative chromatography step, since ion-exchange columns have a large sample capacity and a good resolving power. Molecular exclusion chromatography is usually best reserved for later in the procedure since its sample capacity and resolving power are both relatively limited.

## 5.1 Principles of chromatography

The word "chromatography" means "writing with colour" and refers to the early observations on the separation of dyes by paper chromatography. All chromatographic separations depend upon the differential partition of solutes between two phases, a mobile phase and a stationary phase[1]. Such partition between two phases is described by the so-called partition coefficient or distribution coefficient.

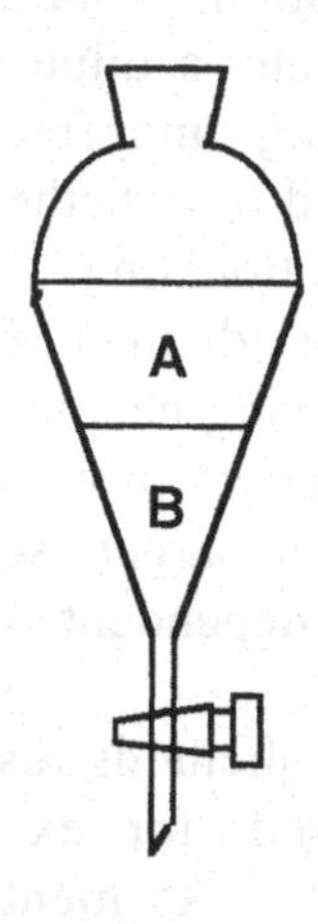


*Figure 51.* Distribution of a solute between phases in a separating funnel.

Students may recall from Chemistry classes how a dyestuff, for example, will distribute itself between two non-miscible liquid phases in a separating funnel. For any two solvents at a constant temperature, the distribution coefficient ($K_d$) is a constant and can be defined as:-

$$K_d = \frac{\text{concentration of solute in A}}{\text{concentration of solute in B}}.$$

The distribution of a solute is not limited to two liquid phases and the distribution coefficient may describe the distribution between any two phases, such as liquid/solid or gas/liquid phases. In chromatography there is always a distribution between two such phases, one kept stationary while the other - the mobile phase - flows over or through it. The stationary phase can therefore be a solid, a liquid, or a solid coated with a liquid. Since it must be fluid, the mobile phase must be either a gas or a liquid.

*Table 2.* Forms of chromatography

| Stationary phase | Mobile phase | Distribution mechanism | Name |
|---|---|---|---|
| solid | liquid | adsorption | Adsorption chromatog. |
| liquid | liquid | partition | Paper chromatography,<br>Counter-current distribution |
| solid | liquid | ion-exchange | Ion-exchange chromatog. |
| liquid (in gel) | liquid | molecular exclusion | MEC |
| immobilised biomolecule | liquid | bio-affinity | Affinity chromatography |
| liquid | gas | partition | GLC |

The mechanism of distribution may not always be simple partition, as in a separating funnel. Examples of the different forms of chromatography are shown in Table 2.

In the case of chromatography, the distribution coefficient is defined as:-

$$K_d = \frac{\text{concentration of solute in stationary phase}}{\text{concentration of solute in mobile phase}}$$

$$= \frac{\text{wt solute in stationary phase/volume of stationary phase}}{\text{wt solute in mobile phase/volume of mobile phase}}$$

$$= \frac{\text{wt in stationary phase}}{\text{wt in mobile phase}} x \frac{\text{volume of mobile phase}}{\text{volume of stationary phase}}$$

$$= k\beta$$

Where $k$ is called the partition ratio (or capacity ratio) and $\beta$ is the phase ratio.

In chromatography the stationary phase is typically packed into a tubular column and the mobile phase flows through the packed column. There is continual equilibration of solutes between the mobile and stationary phases, and that length of column where there is effectively one equilibration - such as would occur in a separating funnel - is called a "theoretical plate". This terminology is derived from fractional distillation of volatile solvents. Since chromatography columns are usually vertically orientated, the length of column in which one equilibration effectively occurs is called the "height equivalent to a theoretical plate", abbreviated HETP. The HETP is more of a concept than a reality, however, because equilibration is actually continuous, not discrete.

The principle of chromatography may be considered by imagining the column to consist of a stack of theoretical plates and, for clarity, the mobile phase may be considered on one side and the stationary phase on the other (*Figure 52*). A column packed with a bead-form stationary phase (A) may be considered as consisting of a vertical stack of theoretical plates (B) in each of which an equilibration between the mobile phase and the stationary phase takes place. Subsequent movement of the mobile phase will displace the mobile "half" of each equilibrated pair downwards, forming new pairs, initially not in equilibrium, but which will equilibrate before, in turn, being displaced.

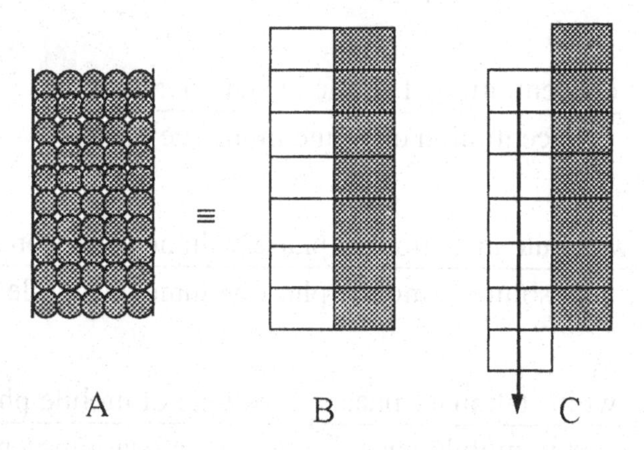


*Figure 52.* A representation of the mechanism of chromatography.

This representation can be used to illustrate the principle of chromatography, as in the tutorial exercise shown in *Figure 53*.

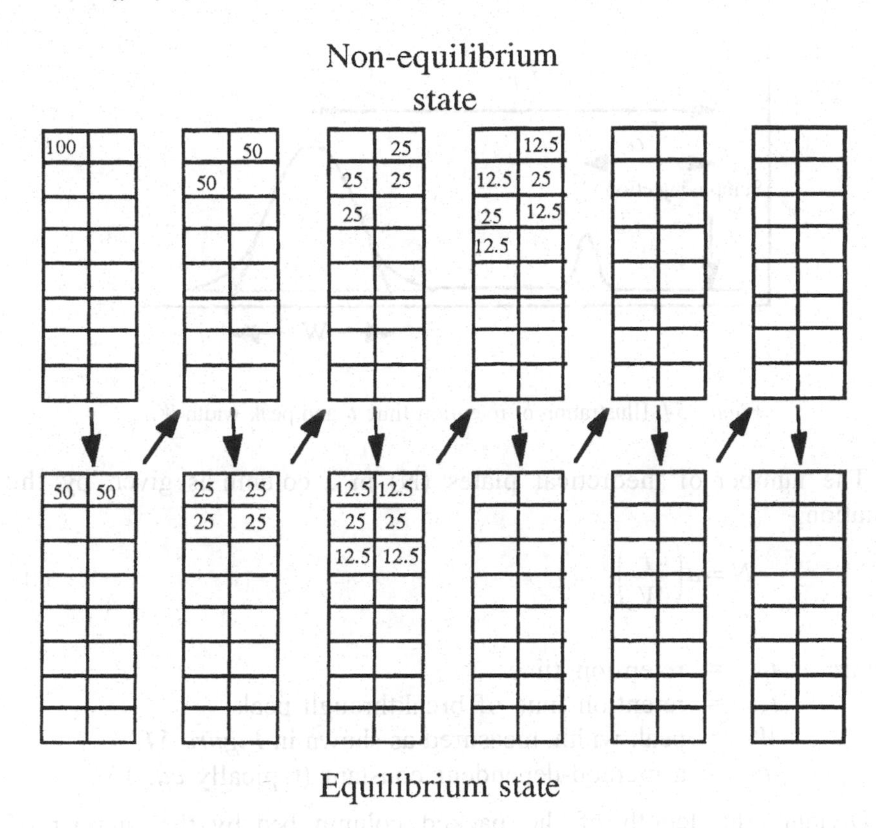


*Figure 53.* A tutorial illustrating the principle of chromatography.

One hundred units of solute are injected into the mobile phase of the column (*Figure 53*, top left). This then equilibrates with the stationary phase (bottom left) - assume a partition ratio of 1 in this case. Movement of the mobile phase carries the solute downwards to a new area of the column (top, second from left), where equilibrium again occurs (bottom, second from left). To see if you have grasped the concept, try to fill in the remainder of the numbers, until the right hand columns are filled in. Note the movement of the "peak", relative to the mobile phase, and note how the peak spreads out.

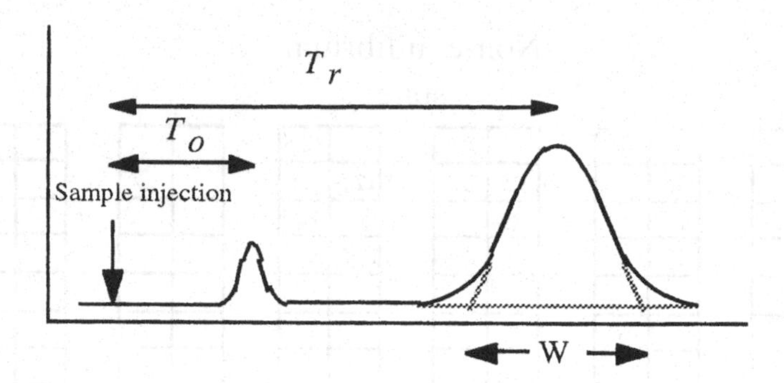


*Figure 54.* Illustration of retention time $t_r$ and peak width *W*.

The number of theoretical plates (N) in a column is given by the equation:-

$$N = a\left(\frac{tr}{W}\right)^2 \qquad 5.1$$

Where $t_r$ = retention time
$t_o$ = retention time of breakthrough peak
$W$ = peak width, measured as shown in *Figure 54*.
$a$ = a method-dependent constant (typically *ca.* 16).

Dividing the length of the packed column bed by the number of theoretical plates gives the HETP. Note that the larger the value of N, the smaller the HETP value and the more efficient the column. For a given retention time, Eqn 4.1 indicates that an efficient column (where N is large) will give peaks of smaller width than an inefficient column.

**Resolution of peaks.**

The resolution (R), which describes how well any two peaks are separated, is described by the equation:-

$$R = \frac{t_{r2} - t_{r1}}{\left(\frac{W_2 + W_1}{2}\right)} \qquad 5.2$$

From this it will be seen that the narrower the peak (i.e. the higher the value of $N$), the better the resolution will be.

The magnitude of the HETP, which should be as small as possible, is influenced by:-

- the particle size of the stationary phase, and,
- the flow rate of the mobile phase.

### 5.1.1 The effect of particle size

Reconsider the equilibration of a solute between two phases in a separating funnel. How quickly equilibrium is achieved will depend upon:-

- diffusion across the boundary between A and B, which is proportional to the surface area of the boundary, and,
- diffusion within A and B to the boundary; the time taken depending upon the distance from the boundary.

For a minimal time to equilibrium, therefore, the boundary surface area should be maximised and the distance of any part of the solutions, A and B, from the boundary should be minimised. This could be achieved by using a separating funnel of unusual design as shown in *Figure 55*.

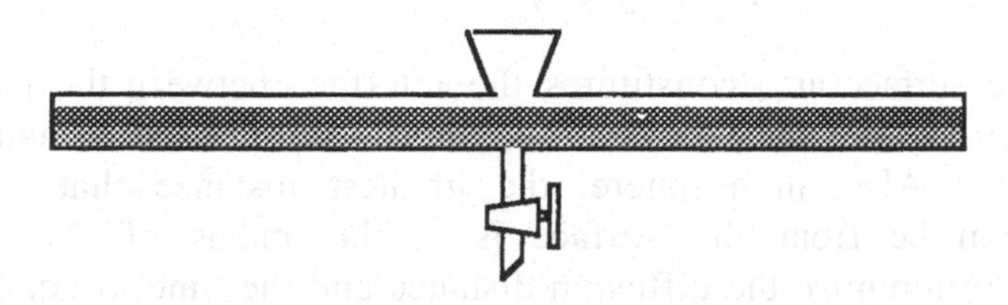

*Figure 55.* A hypothetical separating funnel for rapid equilibration.

However, the more conventional way of speeding up the attainment of equilibrium is by shaking the separating funnel, so that the solutions are well mixed. The two phases remain separate but one solution will be dispersed in the other, usually in the form of small spheres.

The volume of a single sphere is given by:-

$$\text{Volume} = \frac{4}{3}\pi r^3 \qquad 5.3$$

If the total volume of the dispersed phase is $V$, then $V$ will be dispersed into $N$ spheres, where:-

$$N = \frac{V}{\frac{4}{3}\pi r^3} = \frac{3V}{4\pi r^3} \quad 5.4$$

i.e. the number of spheres is inversely proportional to $r^3$.

The surface area of a single sphere is given by:-

$$\text{Surface area} = 4\pi r^2 \quad 5.5$$

Therefore, surface area of "N" spheres

$$= 4\pi r^2 . \frac{3}{4\pi} . \frac{V}{r^3}$$

$$= \frac{3V}{r} \quad 5.6$$

i.e. as the radius of the spheres $r$ gets smaller, the total surface area gets larger.

Since the surface area constitutes the interface between the phases, a small value of $r$ will ensure a maximal interface area and rapid equilibration. Also, in a sphere, the greatest distance that a solute molecule can be from the surface is $r$, the radius of the sphere. Therefore, to minimise the diffusion distance and the time to equilibrium, $r$ should be minimal. The largest possible distance to the surface for a molecule outside of the packing material also decreases as $r$ decreases.

Shaking a separating funnel vigorously is an effective way of making small spheres and hence of rapidly equilibrating the phases. Similarly, for rapid equilibration, the best size for the spherical particles of a chromatography resin is "as small as possible". For even packing and good flow characteristics, the resin particles should also be of uniform size (see equation 3.20).

As $r$ decreases, however, the total surface area increases and so the resistance to the flow of the mobile phase also increases. Very small particles, therefore, dictate the use of high pressure pumps - hence HPLC (high pressure liquid chromatography). To withstand the high pressure, the particles of the stationary phase must be rigid and strong and so silica is commonly used.

### 5.1.2 The effect of the mobile phase flow rate

The effect of the flow rate of the mobile phase is expressed by the so-called Van Deemter equation[2]:-

$$HETP = A + \frac{B}{F} + CF \qquad 5.7$$

Where, $HETP$ = height equivalent to a theoretical plate
$F$ = mobile phase flow rate
$A$ = eddy diffusion (independent of F)
$B$ = molecular diffusion (increases as F decreases)
$C$ = resistance to mass transfer (i.e. smearing) (increases as F increases)

An "eddy" is a swirl in a liquid. "Eddy" diffusion refers to the fact that the mobile phase has to follow a tortuous path around the resin particles, inevitably resulting in some mixing and consequent dilution of a solute peak.

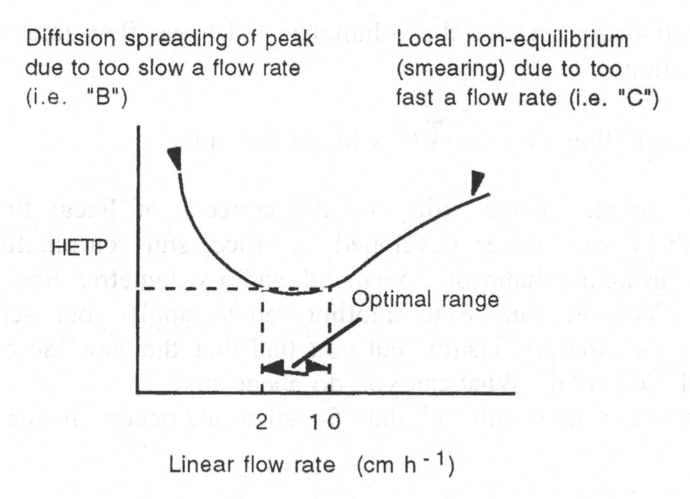


*Figure 56.* A schematic plot of the van Deemter equation.

If there was no flow in the column, the solute peak would spread with time due to diffusion of the solute molecules, from an area where they are in high concentration to an area where their concentration is less. Similarly, when the flow rate is too slow, the peak will have ample time to spread due to diffusion. On the other hand, if the flow is too fast, solute molecules in the mobile phase will pass the stationary phase without properly equilibrating with it, resulting in peak broadening due to "smearing". The optimal linear flow rate for typical low-pressure, molecular exclusion chromatography lies in the range of 2-10 cm $h^{-1}$.

#### 5.1.2.1 The relationship between linear and volumetric flow rates.

The easiest way of measuring the mobile phase flow rate is to collect the effluent stream in a measuring cylinder and measure the amount collected in a given time interval. The result will be the volumetric flow rate, which can be expressed in ml $h^{-1}$. For the chromatographic process, however, the important point is not the volumetric flow rate *per se* but how fast the mobile phase flows past the stationary phase.

A moment's reflection will reveal that a given volumetric flow rate will give very different chromatographic conditions in a thin column compared to a fat column as the mobile phase will flow past the stationary phase faster in the thin column than in the fat column. To make the chromatographic conditions the same in columns of any diameter, the flow rate can be expressed as the linear flow rate, with units of cm $h^{-1}$.

The relationship between the volumetric and linear flow rates is given by the equation:-

$$\text{Volumetric flow rate} = \pi r^2 \times \text{linear flow rate.} \qquad 5.8$$

As an example of the utility of the concept of linear flow rate, imagine that you have developed a successful chromatographic separation, using a column of 2.5 cm i.d. and a volumetric flow rate of 50 ml $h^{-1}$. You then move to another lab to apply your separation method, as a temporary visitor, but you find that the new lab only has columns of 2.0 cm i.d. What can you do about this?

First calculate the length "$\ell$" that 50 ml would occupy in the 2.5 cm column:-

$$50 = \pi r^2 \ell$$

Therefore, $\ell = \dfrac{50}{\pi r^2}$

Hence, $\ell = \dfrac{50}{\pi . 1.25^2}$

$= 10.186$ cm

The linear flow rate in the 2.5 cm i.d. column is thus 10.186 cm $h^{-1}$. From this, calculate the volumetric flow ("x") in the 2.0 cm i.d. column that would give the same linear flow rate, i.e.;-

$x = \pi r^2 \times 10.186$

And for $r = 1$ cm,

$x = 32$ ml $h^{-1}$

Thus the required volumetric flow rate in the 2.0 cm column is 32 cm $h^{-1}$. The 2.0 cm i.d. column could be operated at the same length as the 2.5 cm i.d. column, but at the reduced volumetric flow rate of 32 cm $h^{-1}$.

## 5.2 Equipment required for low pressure liquid chromatography

### 5.2.1 The column

A *sine qua non* for column chromatography is the column. Basically this consists of a glass tube with adapters - preferably at either end - to spread the liquid flow from the thin bore input tubing out to the relatively large bore of the column and back in to the thin bore of the output tubing (*Figure 57*). The column packing is supported on a sieve of some sort and a key element in the efficiency of the column is the "dead" volume between this sieve and the output tubing, which should be as small as possible. The purpose of the column is to effect a separation of different types of solute molecules and the whole purpose is defeated if the separated molecules are allowed to remix, due to the dead volume being too large.

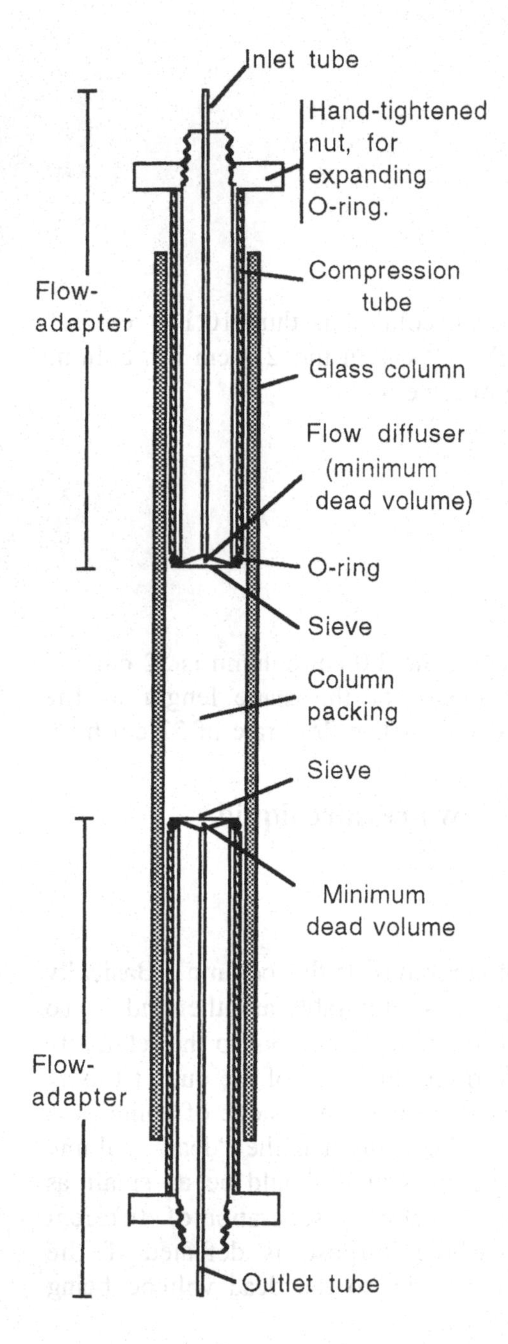


*Figure 57.* Schematic cross-section of a chromatography column.

Moveable flow-adapters enable different column bed volumes to be used. It is useful to have such flow adapters on both ends of the column. After packing and equilibration of the column bed, the adapter on the inlet side can be adjusted to be in contact with the upper surface of the packed resin bed. This is advantageous in that it facilitates sample application, as the sample can be simply introduced through the inlet tubing. It is also necessary if any form of gradient elution is to be used. In the absence of an upper flow adapter, the incoming buffer will mix in the dead volume above the packed resin bed and it will be impossible to get smooth, reproducible, gradient conditions.

The ratio between the internal diameter (i.d.) of the column and its length, the so-called "aspect ratio", differs depending upon the application of the column. Generally, where adsorption occurs, such as in ion-exchange or affinity chromatography, columns with an aspect ratio of *ca.* 1:10 or less are used, whereas for molecular exclusion chromatography aspect ratios of about 1:50 are used.

Low pressure liquid chromatography columns typically consist of a uniform bore, thick walled, glass tube, with plastic adapters etc. Glass is favoured because it is chemically stable, transparent, and has good thermal conductivity. Obviously, all materials used to make columns must be chemically stable to buffers etc. and must not react with sample proteins. Proteins do adsorb to glass to some extent and this can be prevented by silanising the column before use, though this is only warranted for the most critical work.

## 5.2.2 Moving the mobile phase

The chromatographic process requires movement of the mobile phase and the simplest way of effecting this is by siphoning the buffer from a reservoir which is elevated above the end of the outlet tube from the column (*Figure 58*).

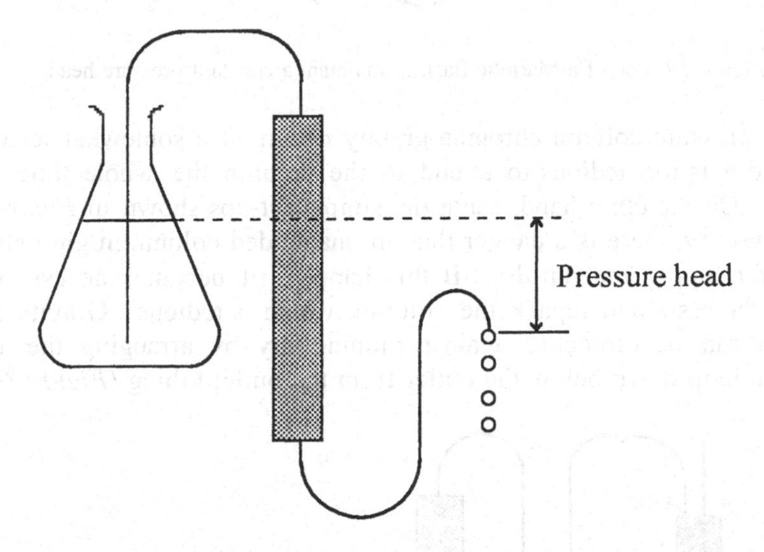


*Figure 58*. Simple chromatography, using a siphon to generate the mobile-phase flow.

The difference in potential energy (height) between the surface of the liquid in the buffer reservoir and the end of the outlet tube constitutes the "pressure head" which will cause the mobile phase to flow. A problem with this simple set-up is that, as the level of liquid in the reservoir drops, the pressure head will get smaller and the flow rate will decline. To keep the pressure head constant a so-called "Marriotte flask" may be used (*Figure 59*).

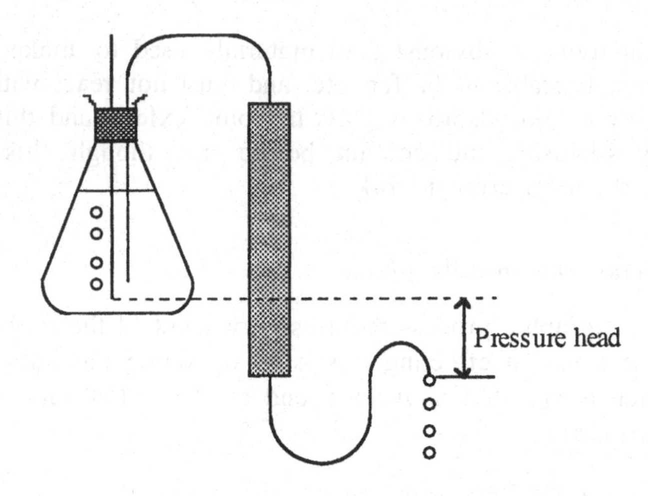


*Figure 59.* Use of a Mariotte flask to maintain a constant pressure head.

Low-pressure column chromatography occurs at a somewhat leisurely pace and it is too tedious to attend to the column the whole time it is running. On the other hand, using the simple set-ups shown in *Figure 58* and *Figure 59*, there is a danger that an unattended column might exhaust the buffer supply and run dry. If this happens it becomes necessary to remove the resin and repack the column, which is tedious. Gravity-flow columns can be protected against running dry by arranging the inlet tubing to loop down below the outlet from the outlet tubing (*Figure 60*).

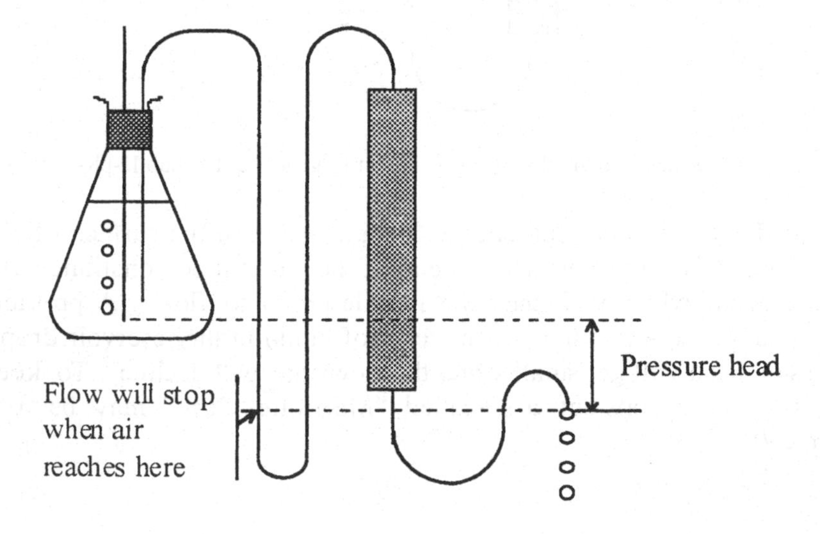


*Figure 60.* A run-dry protection loop on a gravity-flow column.

Note that gravity-flow columns can be operated with the flow going either downwards (*Figure 58→Figure 60*) or upwards through the column (*Figure 61*). An advantage of upward flow is that it is easier to arrange the system so that it will not run dry. Upwards flow is recommended with very soft gels, such as Sephadex G-200, which otherwise tend to be crushed by the combined effects of gravity and the buffer flow. With ascending flow, the flow supports some of the weight of the gel.

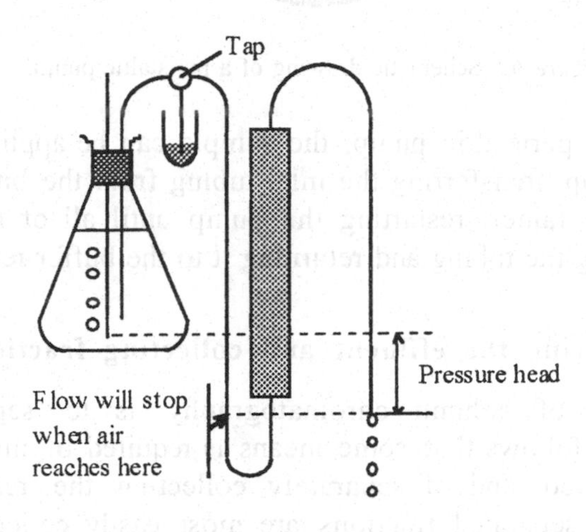


*Figure 61*. Ascending flow, with run-dry protection, and a tap for sample application.

With a gravity flow system the sample is most easily applied using a three-way tap on the inlet side of the column, as shown in *Figure 61*.

Better than a Marriotte flask, if the budget allows, is a peristaltic pump. "Peristalsis" refers to the rhythmic, wave-like, contractions that pump the gut contents along the digestive tract. By analogy, a peristaltic pump is one in which a flexible silicone tube is pinched by a roller which runs along the tube, pumping the tube contents in the same direction as it does so (*Figure 62*). A peristaltic pump gives a smooth, almost pulse-free, flow.

The advantage in using a pump is that it gives more precise flow control and greater freedom in the chromatography lay-out, i.e. the buffer reservoir does not have to be higher than the column outlet. To prevent the column running dry during unattended operation when using a peristaltic pump, a timer switch is required. The timer can be arranged to switch off the pump - and any other associated apparatus - after a pre-set time. Note that because the peristaltic pump rollers pinch the silicone tube, there can be no flow of liquid through the pump when it is switched off.

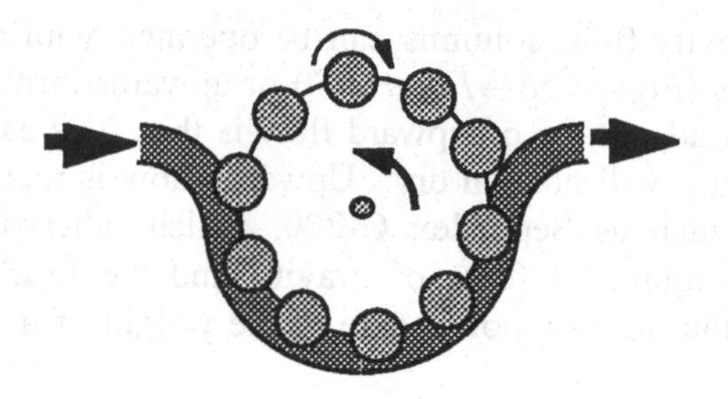

*Figure 62.* Schematic drawing of a peristaltic pump.

When using a peristaltic pump, the sample can be applied simply by stopping the pump, transferring the inlet tubing from the buffer reservoir to the sample container, restarting the pump until all of the sample is sucked up, wiping the tubing and returning it to the buffer reservoir.

### 5.2.3 Monitoring the effluent and collecting fractions.

The purpose of column chromatography is to separate solute molecules and it follows that some means is required of monitoring the separation achieved, and of separately collecting the resolved solute molecules. The separated fractions are most easily collected using an automatic fraction collector. Fraction collectors are available in different versions that collect the effluent stream in fractions on the basis of time, volume, or number of drops. Time-based fraction collectors are the simplest and most economical and, if the mobile phase flow rate is accurately controlled with a pump, the fractions collected will be of equal volume.

The simplest means of monitoring the separation of proteins is to collect the column effluent for the whole run in a convenient number of fractions - say, 100 - and to measure the $A_{280}$ of each fraction in a spectrophotometer. The results can be used to construct a so-called "elution profile", in which $A_{280}$ is plotted against the elution volume.

Such manual reading of the elution profile is inexpensive in capital terms, but it consumes operator time and, perhaps more importantly, some detailed information is lost. A better way, again if the budget can afford it, is to use a flow-through UV-monitor, that continuously reads the absorbance of the effluent stream at 280 nm. Such a monitor is plumbed into the effluent line, between the column and the fraction collector. Besides a power source, it requires two other electrical connections; an output to a recorder and an event-marker connection between the fraction collector and the recorder. Each time the faction collector changes tubes, it sends a pulse to the recorder so that the event - the tube change - is recorded. In this way it becomes possible to subsequently correlate the recorder trace of $A_{280}$ with the collected fractions, so that fractions corresponding to the required peaks can be harvested. A flow-through UV-monitor constitutes a time-saving automatic system which also captures the fine detail of the elution profile.

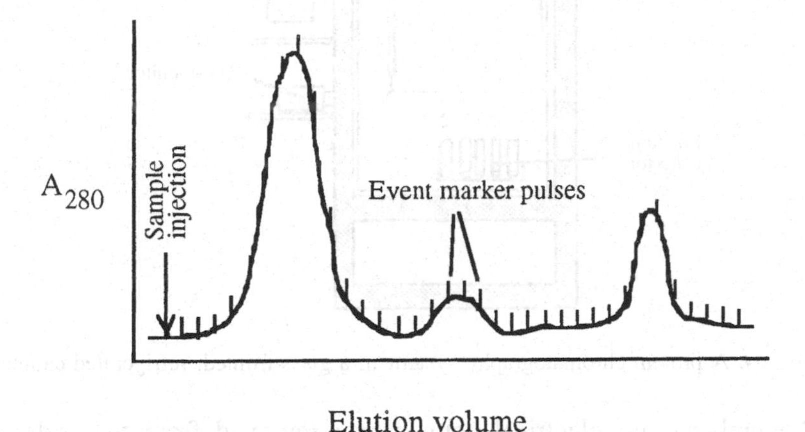


*Figure 63.* A typical elution profile of $A_{280}$ *vs* elution volume, with event marker pulses.

### 5.2.4 Refrigeration

Proteins are structurally labile and are susceptible to microbial degradation. For these reasons, wherever possible, protein solutions are maintained at low temperature and preservatives are added to the buffers. Denaturation and degradation are both minimised by keeping the proteins cold and protein separations are therefore usually carried out at about 4°C.

Chromatography is usually done in coldrooms, but working in coldrooms is miserable and unhealthy. A better and more versatile system is to have a refrigerated cabinet with some components of the chromatography set-up being kept at 4°C and others at room temperature (*Figure 64*). In *Figure 64*, items labelled on the left are within the cabinet at 4°C and those labelled on the right are at room temperature. Electrical apparatus kept in a coldroom or fridge can be damaged by condensation of moisture onto its circuits. For this reason it is best to keep as much as possible at room temperature. The fraction collector must, however, be kept in the fridge.

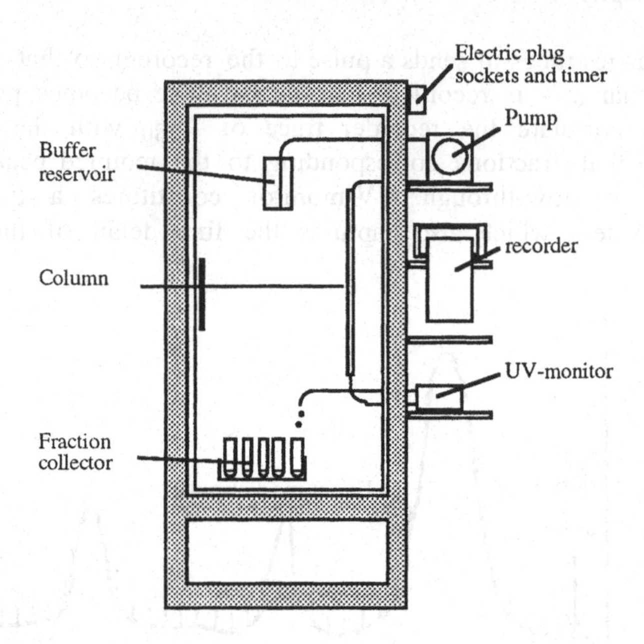


*Figure 64.* A protein chromatography system in a glass-fronted, refrigerated cabinet.

Remember if any electrical item is ever removed from the fridge for servicing, it must be allowed to warm up to room temperature and all moisture must be allowed to dry off before it is switched on. If this is not done, short-circuits caused by condensed moisture may burn out the electronics. Apparatus may be sealed in a plastic bag before removal from the fridge to prevent condensation of moisture on the circuits.

## 5.3 Ion-exchange chromatography (IEC)

Consider the situation where a single anion type, say $CH_3COO^-$, and a number of cation types, say $Na^+$, $K^+$ and $Protein^+$, exist together in an aqueous solution, where the aqueous component will additionally contribute $OH^-$ and $H^+$ ions. These ions will establish a dynamic equilibrium, comprised of a number of sub-equilibria, e.g.

$$CH_3COO^- + Na^+ \leftrightarrow CH_3COONa$$

$$CH_3COO^- + K^+ \leftrightarrow CH_3COOK$$

$$CH_3COO^- + Protein^+ \leftrightarrow CH_3COOProtein$$

$$CH_3COO^- + H^+ \leftrightarrow CH_3COOH$$

The overall equilibrium condition will be determined by the values of the respective dissociation constants:-

$$K_{(CH_3COONa)} = \frac{[CH_3COO^-].[Na^+]}{[CH_3COONa]}$$

$$K_{(CH_3COOK)} = \frac{[CH_3COO^-].[K^+]}{[CH_3COOK]}$$

$$K_{(CH_3COOProtein)} = \frac{[CH_3COO^-].[Protein^+]}{[CH_3COOProtein]}$$

$$K_{(CH_3COOH)} = \frac{[CH_3COO^-].[H^+]}{[CH_3COOH]}$$

The overall equilibrium state is a result of:-

- intrinsic affinities between ions (expressed in the dissociation constants), and,
- competition between ions (a function of the relative concentrations of the ions).

To develop a chromatography system, the $CH_3COO^-$ ion, in the form of a carboxymethyl group ($-CH_2COO^-$), could be covalently attached to the stationary phase and the dissociated cations allowed to move with the mobile phase; nett electrical neutrality being maintained, however. With an immobilised anion, this would constitute a cation exchange system. Conversely, immobilisation of a cation would constitute an anion exchange system.

If the above ions were applied to the chromatography system as a sample "plug" and the system was subsequently eluted with a buffer, say lithium citrate, the differential affinities of the ions for the stationary anion would be manifest as differential rates of migration through the column. The ions would migrate at relative rates proportional to their respective dissociation constants. Their manifest affinity, and thus their absolute rates of migration, would depend upon the *competition* that they encountered from the buffer cation, $Li^+$ in this example. With increasing $Li^+$ concentration, the sample cations would face increasing competition in their association with the immobilised anion and so would be increasingly dissociated, resulting in an increase in their rate of migration through the column. In the case of proteins, the dissociation constant is affected by pH, so elution can also be effected by a change in pH.

### 5.3.1 Ion-exchange "resins"

The term "resin" comes from early polystyrene-based ion-exchangers which had a translucent yellow appearance, like the resin exudates from pine trees. The term has stuck, although modern ion-exchangers used for protein separations are generally opaque and white.

All ion-exchange resins are comprised of a matrix to which are attached ionic substituent groups. For low pressure chromatography of proteins, the matrix is often comprised of a hydrophilic biopolymer, such as cellulose, Sephadex™, or agarose. These materials cannot withstand high pressures and for medium to high pressure liquid chromatography, the trend is towards silica-based resins, or synthetics such as Trisacryl™.

Cellulose is a polymer of β-D-glucose units, linked with 1→4 bonds. It is relatively inexpensive and provides good flow properties, but large interstitial spaces lead to relatively poor resolution. Sephadex consists of dextran chains, comprised of 1→6 linked dextrose (glucose) residues, cross-linked with epichlorhydrin (see *Figure 75*). The name "Sephadex" is a contraction of the words "*se*parating", "*Pha*rmacia" and "*dex*tran". It is sold in the form of dry xerogels which absorb water and swell into hydrated spherical particles. Substituted Sephadex ion-exchangers give good resolution but they are subject to marked volume changes with changes in buffer ionic strength. This is a disadvantage as it is difficult to apply an accurate salt gradient to a shrinking gel, and it may become necessary to re-pack the column after only a few runs.

Agarose is the neutral polysaccharide component of agar, an extract of kelp, which is a type of seaweed. It is a linear polysaccharide composed of alternating residues of D-galactose and 3,6-anhydro-L-galactose, linked by β1→4 and α1→3 bonds.

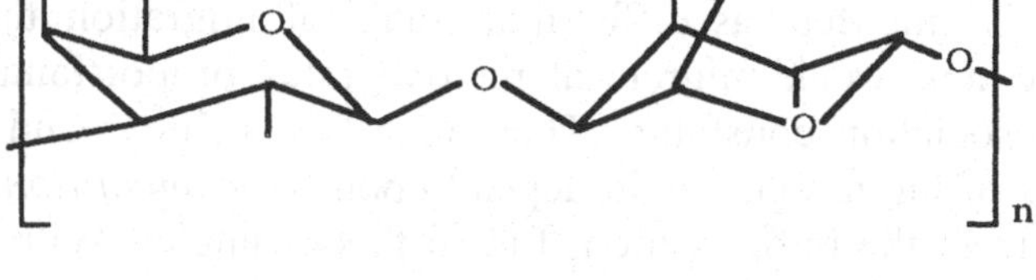

*Figure 65.* The structure of agarose.

Agarose is freely soluble in water at 100°C and upon cooling forms an exceptionally strong, so-called macroreticular gel, with large pores (*Figure 66*).

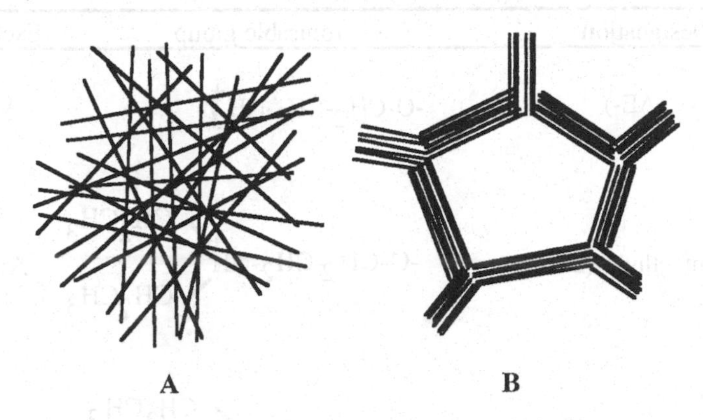


*Figure 66.* Comparison of micro- and macroreticular gels.

Sephadex (*Figure 66*A) is an example of a microreticular gel. In a macroreticular gel (*Figure 66*B), e.g. agarose, the gel fibres align into bundles resulting in a much stronger gel and a larger pore size at a given gel concentration. The sketch in *Figure 66* is a 2-D representation, but gels actually form 3-D labyrinths.

The macroreticular structure of agarose makes it very suitable as a matrix for ion-exchangers as proteins have easy access to the gel interior, so that in effect the gel has a very large surface area to which ionic substituent groups may be attached. The macroreticular structure is also mechanically strong, so that substituted agarose ion-exchangers do not shrink or swell with changes in buffer ionic strength. The gel structure of agarose is maintained by non-covalent bonds and agarose gels cannot be dried and reconstituted. They are consequently supplied in the form of a slurry. They also cannot be boiled or autoclaved, as they simply melt to a sol at high temperatures.

Common substituent groups are shown in Table 3. The common weak base anion exchanger group is DEAE- and the common weak acid cation exchanger group is CM-. The pH range over which these groups are ionised is shown in *Figure 67*.

*Table 3.* Some common ion-exchange substituent groups.

| Designation | Ionisable group | Exchanger |
|---|---|---|
| Aminoethyl- (AE-) | $-O-CH_2-CH_2-NH_3^+$ | Anion |
| Diethylaminoethyl- (DEAE-) (Weakly basic) | $-O-CH_2-CH_2-\overset{+}{N}H(CH_2CH_3)_2$ | Anion |
| Triethylaminoethyl- (TEAE-) | $-O-CH_2-CH_2-\overset{+}{N}(CH_2CH_3)_3$ | Anion |
| Trimethylaminomethyl- (Q-) (Strongly basic) | $-O-CH_2-\overset{+}{N}(CH_3)_3$ | Anion |
| Carboxymethyl- (CM-) (Weakly acidic) | $-O-CH_2-COO^-$ | Cation |
| Sulfomethyl- (S-) (Strongly acidic) | $-O-CH_2-SO_3^-$ | Cation |

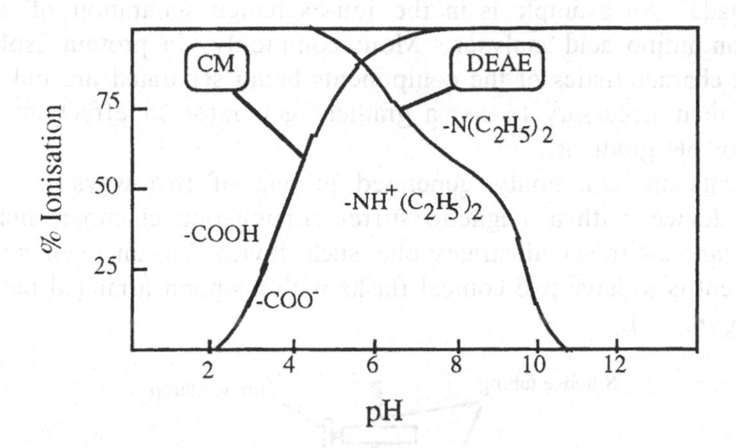


*Figure 67.* Ionisation characteristics of CM- and DEAE- substituent groups.

It can be seen from *Figure 67* that exchangers comprised of weak acid or weak base substituent groups are not completely ionised at most pH values of interest and, perhaps a greater drawback, the degree of ionisation changes with pH in the useful pH range. Exchangers based on strong acids or bases, by contrast, are completely ionised over a much larger pH range, so their degree of ionisation is less subject to change with pH (*Figure 68*).

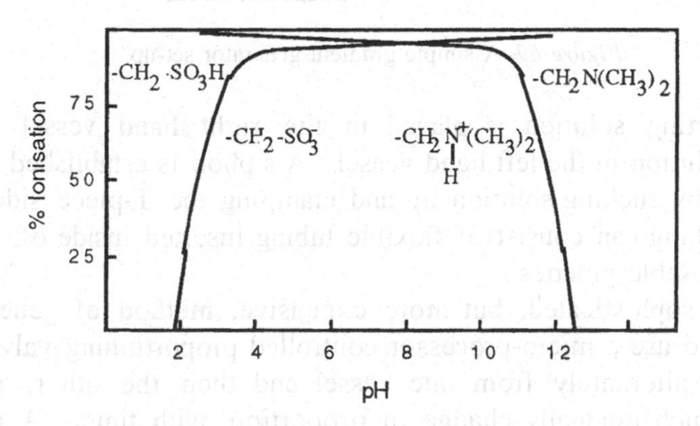


*Figure 68.* Ionisation of strong acid and strong base ion-exchange substituent groups.

### 5.3.2 Gradient generators

Ion-exchange chromatography often requires elution of the bound proteins by a change in either ionic strength, pH, or both. If the system being separated is well characterised, then appropriate stepwise changes can be made. An example is in the ion-exchange separation of amino acids in an amino acid analyser. More commonly, in protein isolation the exact characteristics of the components being separated are unknown and it is then necessary to use a gradient generator to effect an ionic strength or pH gradient.

Gradients are commonly generated in one of two ways. A two-chamber device, with a magnetic stirrer stirring one chamber, may be used. *Figure 24* (p54) illustrates one such device, but an even simpler arrangement is to have two conical flasks with a siphon arranged between them (*Figure 69*).

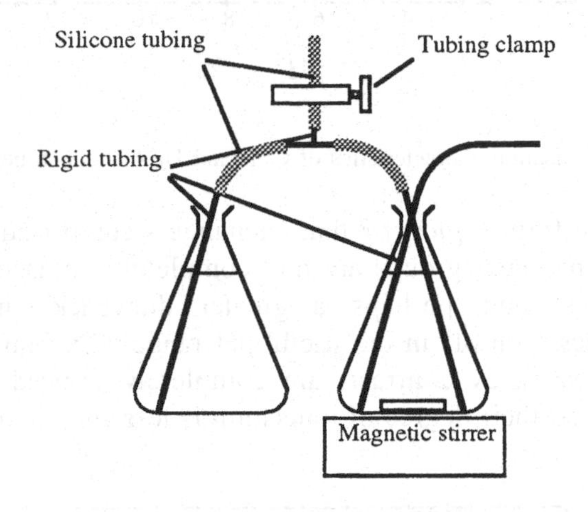


*Figure 69.* A simple gradient generator set-up.

The starting solution is placed in the right hand vessel and the finishing solution in the left hand vessel. A siphon is established between the vessels by sucking solution up and clamping the T-piece side tubing. The rigid tubing can consist of flexible tubing inserted inside of, for e.g., plastic disposable pipettes.

A more sophisticated, but more expensive, method of generating a gradient is to use a micro-processor controlled proportioning valve which draws liquid alternately from one vessel and then the other, in small amounts which gradually change in proportion with time. A mixer is placed in the line, downstream of the proportioning valve, to change the small stepwise changes in buffer composition into a gradual and continuous change. Gradient generators based on proportioning valves are common components of complete chromatography systems, commercially available from a number of manufacturers.

### 5.3.3 Choosing the pH

One of the decisions that has to be made before conducting ion-exchange chromatography is what pH to use. The selection of pH and of the type of ion-exchanger to use may be facilitated by establishing the so-called titration curves of the proteins in the mixture to be separated. An electrophoretic titration curve can be determined by establishing a pH gradient in a gel (see Section 6.10) and conducting an electrophoretic separation (see Section 6.8) at right-angles to the pH gradient (*Figure 70*).

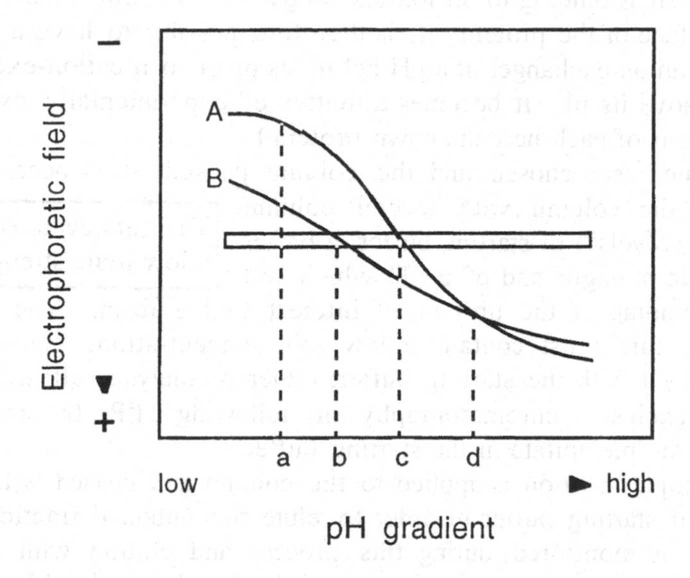


*Figure 70.* Electrophoretic titration curves of proteins.

In *Figure 70*, pH "a" is the pH of maximal charge difference between proteins A and B. At this relatively low pH, both proteins have a positive charge and a cathodic migration. This information suggests that cation-exchange chromatography, conducted at pH "a", would most likely effect the best separation between proteins A and B. By contrast, at pH "d" the proteins have no apparent charge difference. This implies that anion-exchange chromatography would be less successful at resolving A and B, especially at pH "d". Points "b" and "c" represent the pI values of proteins B and A, respectively. The difference in these pI values gives an indication of the separation which may be expected from isoelectric focusing (Section 6.10) or chromatofocusing (5.4).

### 5.3.4 An ion-exchange chromatography run

The choice of the type of exchanger, cation- or anion-exchanger, may be arbitrary, in the absence of any knowledge of the characteristics of the protein of interest, e.g. its pI and/or its pH stability range. As a first approach, it is generally best to choose a pH, within the protein's stability range, where it will be bound to the ion-exchanger. Usually, this means that an anion-exchanger should be used at pH values above the pI of the protein and a cation-exchanger at a pI below the pI. (It must be realised, however, that the pI refers to an overall property of the protein, whereas binding to an ion-exchanger is a function of the charge on one surface of the protein. It is therefore possible to have a protein bind to an anion-exchanger at a pH below its pI or to a cation-exchanger at a pH above its pI. It becomes a matter of experimentally exploring the behaviour of each new unknown protein.)

With the resin chosen and the column packed, it is necessary to equilibrate the column with several column volumes ("colvols") of starting buffer, a buffer of low ionic strength and of a pH which will promote binding of the protein of interest to the resin. The sample protein mixture must contain a low salt concentration, achieved by equilibrating it with the starting buffer; either by dialysis, ultrafiltration, molecular exclusion chromatography or, following TPP, by simply re-dissolving the precipitate in the starting buffer.

> The sample must have a low ionic strength.

The sample solution is applied to the column and chased with about 2 colvols of starting buffer in order to elute the unbound fraction. The $A_{280}$ may be monitored during this process and elution with starting buffer stopped once the $A_{280}$ returns to the baseline. At this point a buffer gradient may be applied. As a first approach, a gradient of increasing ionic strength is the best choice, and is applicable to both cation- and anion-exchange.

The resolution of peaks is a function of the steepness of the eluting gradient. A shallow gradient gives better resolution, but takes more time, so a trade-off must be made. With an unknown system, the best first approach is to use a steep gradient, as this gives a quick assessment of the number of peaks to be expected, and the separation can subsequently be optimised. A suitably steep gradient for a first approach is 0→1 M NaCl in starting buffer, in 3 colvols. For subsequent optimisation the gradient limits or the number of colvols can be altered, to change the gradient slope.

> A shallow gradient gives better resolution, but takes more time

It must be appreciated that the gradient is applied to the inlet side of the column, whereas monitoring of the effluent is done on the outlet side. There is thus a colvol difference between the influent and effluent streams. Consequently, after application of the gradient, it is necessary to elute with at least one colvol of finishing buffer to ensure that the whole gradient itself is eluted and that all peaks which would be eluted by the gradient are, in fact, washed from the column. An example of the reporting of an ion-exchange chromatography run is presented in *Figure 71*.

Finish with 1 colvol of finishing buffer.

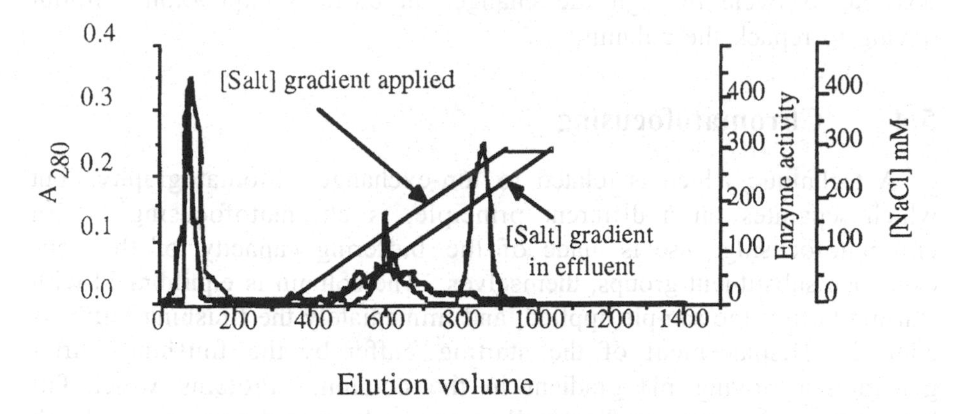


*Figure 71.* An example of ion-exchange chromatography: Purification of cathepsin S, from a TPP fraction from bovine spleen, by chromatography on S-Sepharose. Column, 2.5 cm i.d. x 17 cm; Starting buffer, 20 mM Na-acetate containing 1 mM EDTA and 0.02% $NaN_3$, pH 5.0; Gradient, 0→300 mM NaCl in 600 ml. (Note that the column is eluted with 1 column volume of finishing buffer after the gradient is applied, in order to elute proteins which may have been displaced by the buffer but not yet eluted from the column.) The thick line indicates cathepsin S activity and the thin line is $A_{280}$. (The apparent activity in the break-through peak reflects the fact that the assay is not absolutely specific for cathepsin S).

S-Sepharose is a cation exchanger. If an anion exchanger were used, elution with an ionic strength gradient could be effected in exactly the same way. A different strategy is needed for cation and anion exchangers, however, if a pH gradient is used to elute the bound proteins. In the case of a cation exchanger, a gradient of rising pH is required, whereas with an anion exchanger a gradient of descending pH is required.

After completion of the chromatography run, with either a cation or anion exchanger, firmly bound sample components may be eluted with a high salt concentration, such as 1 M NaCl.

Wash with high [salt] and re-equilibrate with starting buffer

In the case of cation exchangers, high salt concentration may be combined with a high pH, and with anion exchangers the high salt may be combined with a low pH. The extreme pH values used must, of course, be within the stability limits of the ion-exchanger resin. Finally, the column is re-equilibrated with several colvols of starting buffer, in preparation for the next run.

Ion-exchange chromatography thus requires cycling through large changes in ionic strength and possibly also of pH. As mentioned previously, it is an advantage if the resin can withstand the required ionic strength and pH changes, without shrinking or swelling, as this makes it possible to cycle through the changes in buffer composition without having to repack the column.

## 5.4 Chromatofocusing

A technique which is related to ion-exchange chromatography, but which separates on a different principle, is chromatofocusing[3-6]. In chromatofocusing, use is made of the buffering capacity of the ion-exchange substituent groups, themselves. The column is equilibrated with starting buffer, the sample applied, and immediately the finishing buffer is applied. Displacement of the starting buffer by the finishing buffer generates a moving pH gradient in the column. Proteins which fall behind their pI on this gradient will no longer bind to the column and will be swept along faster than the pH gradient. Proteins which move ahead of the pH gradient, by contrast, will bind strongly to the column and will be immobilised until overtaken by the pH gradient. The nett result is that proteins will be eluted from the column at their respective pI values.

In practice it is found that simple displacement of one buffer with another in a conventional exchanger causes too sharp a change in pH. A shallower pH gradient, more suitable for chromatofocusing separations, can be generated by using ampholytes as the eluting buffer, and a substituent group, such as polyethyleneimine, which titrates over a larger pH range. Ampholytes are mixtures of randomly substituted poly amino-poly carboxylic acids. They are also used in isoelectric focusing and in isotachophoresis. This requirement for ampholytes makes chromatofocusing more expensive than normal ion-exchange chromatography.

## 5.5 Molecular exclusion chromatography (MEC)

As shown in *Figure 66*, gels are comprised of a large volume of water immobilised by a relatively small volume of hydrophilic polymer fibres, arranged in a randomly ramified 3-D network. Covalent or non-covalent cross-links between the fibres make the gel insoluble.

In molecular exclusion chromatography, the gel is arranged in the form of small, uniformly-sized spheres ("beads") which are suspended in buffer and packed into a column. In this situation the aqueous solvent may be considered in two parts; that within the gel spheres, which is held stationary, and that between the gel spheres which is free to move.

*Figure 72* is a 2-D representation of part of a gel structure and shows that a smaller particle "A" has access to a larger volume of the immobilised water, as indicated by the shaded area in the left hand figure, compared to that in the right hand figure. The shaded area indicates the accessible water while the inaccessible water is the unshaded area bounded by the lines representing the gel fibres. The proportion of the stationary phase to which a solute molecule has access is thus inversely proportional to its size, i.e. smaller solute molecules (A) can "get at" more of the water within the gel beads, whereas larger particles (B) are able to "get at" less of the immobilised water. Expressed the other way around, we may say that the larger molecules are "excluded" from a larger proportion of the immobilised solvent, and this gives the technique its name.

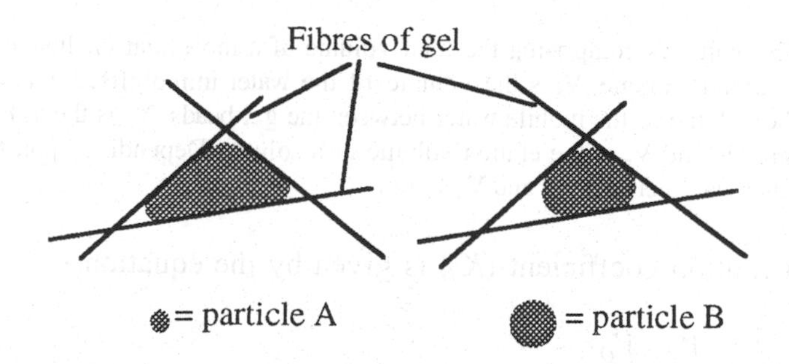


*Figure 72.* Gel volumes accessible to proteins of different sizes.

The differential access of differently-sized solute molecules to the stationary phase is reflected in differences in their distribution coefficients, and consequently in their chromatographic behaviour. Large molecules, which have little access to the stationary phase (reflected in a small value of the distribution coefficient), will elute before smaller molecules which have greater access to the stationary phase (reflected in larger distribution coefficient values), and which will be relatively retarded. The order of elution from a molecular exclusion column is therefore in decreasing order of molecular weight.

*Figure 73* shows a representation of the different volumes which contribute to the total bed volume of a molecular exclusion column. The total volume $V_t$ is made up of the void volume $V_o$, which is the space between the gel beads, plus the volume of the stationary phase $V_s$, which is the immobilised water within the gel beads, plus the volume of the gel-forming polymer strands $V_g$.

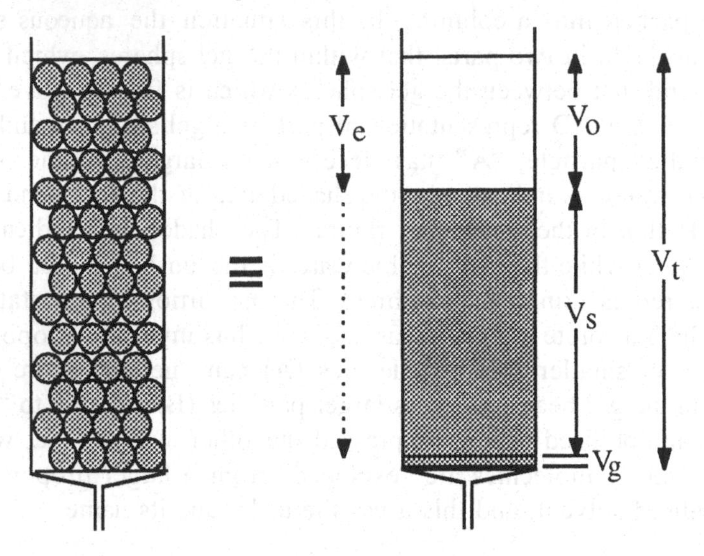


*Figure 73.* The volumes comprising the total volume of a molecular exclusion column. $V_t$ is the total column volume, $V_s$ is the volume of the water immobilised within the gel beads, $V_o$ is the volume of the mobile water between the gel beads, $V_g$ is the volume of the gel polymer strands and $V_e$ is the elution volume of a solute. Depending upon the size of the solute, $V_e$ may vary between $V_o$ and $V_o + V_s$.

The distribution coefficient ($K_d$) is given by the equation:-

$$K_d = \frac{V_e - V_o}{V_s}$$

Where, $V_e$ = elution volume
$V_o$ = void volume (volume of the mobile phase, which is the liquid between the gel spheres)
$V_s$ = volume of the stationary phase.

$K_d$ has the limits of 0 (when $V_e = V_o$), and 1 (when $V_e = V_o + V_s$). A practical problem with the use of $K_d$, is that it is difficult to measure $V_s$ values and so an alternative, the "availability constant", $K_{av}$, is more commonly used. $K_{av}$ is an approximation of $K_d$, because in defining $K_{av}$ it is assumed that $V_g$, the volume of the gel-forming polymer strands, is negligible. $K_{av}$ is defined as the fraction of the stationary phase which is *available* to a given solute, and is described by the equation:-

$$K_{av} = \frac{V_e - V_o}{V_t - V_o}$$

where, $V_e$ = elution volume
$V_o$ = void volume
$V_t$ = total column volume.

For all practical purposes, however, $K_{av}$ may be considered the same as $K_d$: it has the same limits of 0 (when $V_e = V_o$, i.e. for large molecules) and 1 (when $V_e = V_t$, i.e. for small molecules).

Note that all molecules should elute between the limits of $V_o$ and $V_t$. This is a limitation of MEC which distinguishes it from IEF, for example, where the elution volume can be varied greatly and can extend many fold greater than the column volume.

Since $V_t$ is fixed by the dimensions of the packed column bed, it may be imagined that, in MEC, a greater volume for separation of peaks could be created by reducing $V_o$. Could this be achieved by reducing the size of the gel beads?

### 5.5.1 The effect of gel sphere size on $V_o$

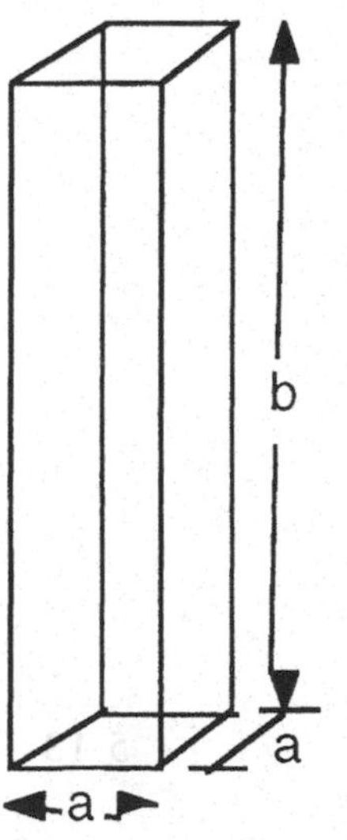


Consider a column of square cross-section with a side $a$ and length $b$, filled with spheres of radius $r$, closely packed in uniform layers. On average, ignoring edge effects, which are small if $r$ is small compared to the column dimensions, the number of spheres ($N_S$) in one layer is given by:-

$$N_s = \left(\frac{a}{2r}\right)^2 \qquad 5.9$$

The vertical separation between layers (d) is:-

$$d = \sqrt{\frac{2}{3}}.2r \qquad 5.10$$

Therefore, the number of layers $N_l$ is;-

$$N_l = \frac{b}{\sqrt{\frac{2}{3}}.2r} \qquad 5.11$$

From Eqns 5.9 and 5.10, the number of spheres in the column $N_{tot}$ is given by:-

$$N_{tot} = \frac{b}{\sqrt{\frac{2}{3}}.2r} \cdot \left(\frac{a}{2r}\right)^2$$

$$= \frac{ba^2}{8r^3\sqrt{\frac{2}{3}}} \qquad 5.12$$

The volume occupied by $N_{tot}$ spheres, $Vsp$ is:-

$$Vsp = \frac{ba^2}{8r^3\sqrt{\frac{2}{3}}} \cdot \frac{4}{3} p r^3$$

$$= ba^2 \cdot \left( \frac{4\pi}{8x3x\sqrt{\frac{2}{3}}} \right)$$

$$= 0.6413\ ba^2$$

Now, $V_t = V_o + Vsp$

Therefore, $V_o = V_t - Vsp$

i.e. $V_o = ba^2 - 0.6413\ ba^2$

Hence, $\mathbf{V_o \approx 0.36 V_t}$ 5.13

Eqn 5.13 implies that the void volume is independent of the size of the gel beads, if these are of uniform size, and is proportional only to the total column volume, being just over one-third of $V_t$. This may strike one as a surprising, and perhaps counter-intuitive, result!

It follows from Eqn 5.13 that in MEC the phase ratio (see p91), which is the volume of the mobile phase $V_o$ divided by the volume of the stationary phase ($V_t$ - $V_o$), is:-

$$\frac{V_o}{V_t - V_o} = \frac{0.36\,V_t}{0.64\,V_t} \approx 0.56$$

Uniformly sized spheres are desirable, because these pack evenly under gravity (see Eqn 3.20, p52), and small spheres are desirable because these give the fastest equilibrations between the mobile and stationary phases (see Eqn 5.6, p96). This raises the question of how the required small, uniformly sized, gel beads may be made.

### 5.5.2 The manufacture of small, uniform, gel spheres

Certain non-polar liquids, such as petroleum ether, are not miscible with water but are less dense than water and will tend to float as a separate phase on top of an aqueous phase. Other non-polar liquids, such as chloroform, are also not miscible with water but are more dense and will separate out as a phase beneath the aqueous phase. It follows that there must exist some mixture of petroleum ether and chloroform which has a density exactly equal to that of the aqueous phase (containing gel-forming polymers). If an aqueous solution is dispersed into such an organic solvent mixture, it would form spherical "droplets" of aqueous phase, suspended in the organic phase[7]. The suspended droplets can be made smaller by vigorous agitation of the solution, and by the addition of an emulsifying agent, which will stabilise the individual droplets and prevent them from coalescing into larger spheres. If a gelling reaction is induced while the aqueous phase is dispersed as small, spherical, droplets this will result in the formation of small, spherical gel particles. The beads formed may be subsequently size-fractionated by sedimentation through a column of water.

### 5.5.3 Determination of *MW* by MEC

The factor having the greatest influence upon the $V_e$ and $K_{av}$ of a protein subjected to MEC is the size (i.e. the radius, $r$) of the molecule which, in the case of globular (roughly spherical) molecules, is related to the molecular weight as follows:-

$$MW = \frac{4}{3}\pi.r^3\rho$$

Therefore,

$$r\,\alpha\,\sqrt[3]{MW}\,.$$

Many attempts have been made to establish relationships between elution volume and molecular weight and to construct calibration curves relating these two factors. Some of these are based upon theoretical models and some are empirical.

An early attempt was that of Laurent and Killander[8], who used a model in which a gel is visualised as being comprised of a random network of rigid fibres. The space between the rigid gel fibres, available to spherical particles, was calculated as a function of the sphere radius. A plot of;

$$\sqrt{-\ln K_{av}} \quad vs \quad \sqrt[3]{MW}$$

gave a sigmoidally-shaped curve. The middle part of this curve was approximately linear and a curve constructed with proteins of known molecular weight could be used to estimate the molecular weights of unknown proteins. Hjertén[9] has used a thermodynamic model to arrive at a plot not too dissimilar to that of Laurent and Killander. Hjertén's plot is;

$$-\ln K_{av} \quad vs \quad (MW)^{\frac{2}{3}}$$

Andrews[10] has proposed an empirical plot of;

$$V_e \quad vs \quad \log MW\,.$$

Again, this gives a sigmoidal curve, with an approximately linear centre section. Although it lacks a theoretical basis, the Andrews plot has the over-riding merit of simplicity. A limitation, however, is that by plotting $V_e$ values it is the particular column that is standardised. If that column runs dry or is repacked, it becomes necessary to construct a new standard curve. This limitation of the Andrews plot has been addressed by Fischer[11] who has proposed plotting;

$$K_{av} \quad vs \quad \log MW$$

Using the Fischer plot it is the particular *gel* that is standardised, not the column, so the standard curve established with one column can be applied to another column packed with the same gel. In view of its simplicity and versatility, the Fischer plot (*Figure 74*) has become the most commonly used method of estimating *MW*s from MEC data.

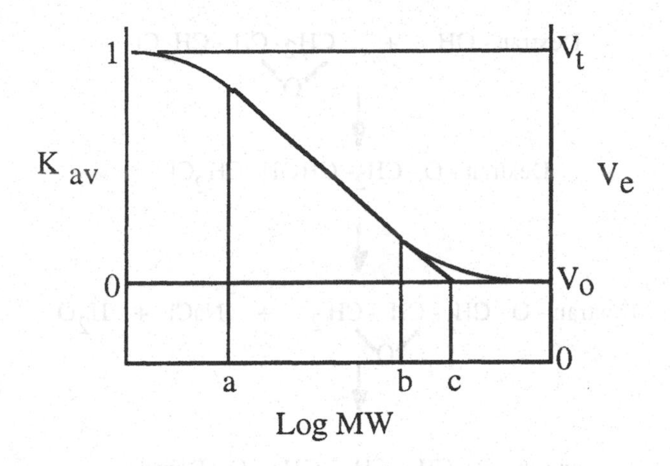


*Figure 74.* A Fischer plot, relating $K_{av}$ to Log *MW*.

In *Figure 74*, the range "a" to "b" represents the effective separating range of the gel. Point "c" gives the *MW* corresponding to the exclusion limit of the gel.

### 5.5.4 Gels used in MEC

A number of different types of gels, made by different manufacturers, are commercially available. In each case the gel is comprised of a network of cross-linked hydrophilic polymers. The nature of the polymer and the type of the cross-link affect the properties of the gel, especially the non-sieving interactions with the sample proteins. These interactions may or may not improve a particular separation and the only way to find out is to try the different gels with the proteins one is trying to separate.

A type of gel which has been popular almost since the advent of MEC is Sephadex, a product of Pharmacia Biotech. (now incorporated with Amersham Biosciences), which consists of dextran (1→6 linked polymer of glucose), crosslinked with epichlorhydrin (*Figure 75*).

Different porosities of Sephadex gels are prepared by varying the degree of cross-linking. The different gels are designated by G-numbers, viz. G-10→G-200. Sephadex G-10 has the smallest pore size and the smallest water regain value, i.e. the ml water taken up per gram of xerogel, and Sephadex G-200 has the largest pores and the largest water regain. Because it consists of a large volume of water immobilised by a relatively small amount of gel, Sephadex G-200 is relatively soft and is best used with ascending elution.

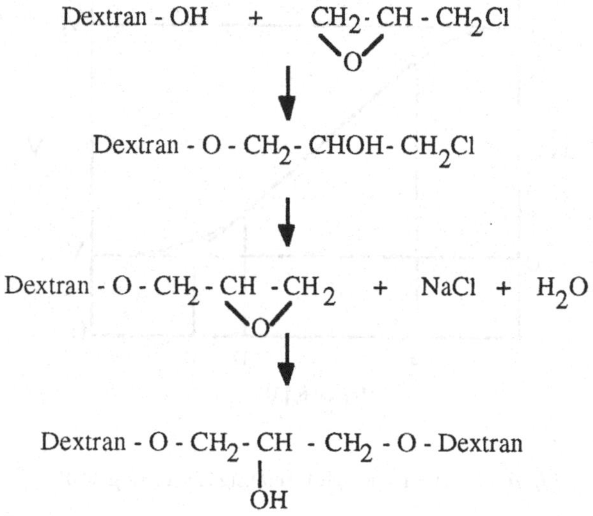

$$\text{Dextran-OH} + \underset{\diagdown O \diagup}{CH_2 - CH} - CH_2Cl$$

$$\downarrow$$

$$\text{Dextran} - O - CH_2 - CHOH - CH_2Cl$$

$$\downarrow$$

$$\text{Dextran} - O - \underset{\diagdown O \diagup}{CH_2 - CH - CH_2} + NaCl + H_2O$$

$$\downarrow$$

$$\text{Dextran} - O - CH_2 - \underset{OH}{\underset{|}{CH}} - CH_2 - O - \text{Dextran}$$

*Figure 75.* The cross-linking of dextran chains to form Sephadex gel.

The cross-links appear to have some hydrophobic character because Sephadex gels will bind small, hydrophobic, molecules with an affinity that is proportional to the degree of cross-linking, i.e. hydrophobic binding is more marked with Sephadex gels with a small G-number. Tryptophan, for example, emerges from Sephadex G-100 at a volume greater than $V_t$, and dyes generally bind tightly to Sephadex G-25.

> $V_e > V_t$ indicates that the sample binds to the gel

**Advantages** of Sephadex are that it is stable over the pH range 2-12, it is biologically inert and, being a permanent gel it is autoclavable, i.e. it does not melt to a sol upon heating. **Disadvantages** are the hydrophobic interaction mentioned above and the fact that, being constituted of biopolymers, Sephadex is biodegradable and is thus subject to microbial degradation. It is thus necessary to include preservatives in the buffers used with Sephadex gels.

PDX, a product of Polydex Biologicals Inc., is also comprised of cross-linked dextran polymers, in bead form. PDX GF 25 separates in the range 1→5 kDa and PDX GF 50 in the range 1.5→30 kDa.

Bio-Gel® is the trade name of Bio-Rad Laboratories of California, for a range of gels for molecular exclusion chromatography. Bio-Gels are produced by co-polymerizing acrylamide (I) with N,N'methylene bisacrylamide (II) to form a cross-linked polyacrylamide.

$$\textbf{I} \quad CH_2{=}CH-CONH_2$$

$$\textbf{II} \quad CH_2{=}CH-\underset{\underset{O}{\|}}{C}-NH-CH_2-NH-\underset{\underset{O}{\|}}{C}-CH=CH_2$$

Again, the porosity is varied by varying the proportion of cross-linking agent (N, N', methylene bisacrylamide), resulting in a range of gels denoted Bio-Gel P-2, P-10, P-20, P-30, P-100, etc. Here the numbers refer approximately to the exclusion limit x $10^{-3}$; thus Bio-Gel P-20 has an exclusion limit of about 20 kDa. The "exclusion limit" refers to the molecular weight of a globular protein which, theoretically, is just completely excluded from the gel (see *Figure 74*). This is, of course, conceptually equivalent to the molecular weight of a globular protein that is just *not* excluded from the gel.

Bio-Gel is chemically inert and is a permanent gel and so can be autoclaved. It also binds some compounds by a hydrophobic mechanism. Bio-Gel, however, is not comprised of biopolymers and so is less subject to microbial degradation.

Agarose is a linear polysaccharide comprised of alternating residues of D–galactose and 3,6–anhydro–L–galactose. Agarose itself forms reversible gels, which melt to the sol at high temperature. The melting and gelling temperatures are different - an example of hysteresis - and commercial grades of agarose are available with high or low melting and gelling temperatures. Agarose in bead form, suitable for MEC, is available under the trade names of Sepharose (Pharmacia Biotech) and Bio-Gel A (Bio-Rad Laboratories), among others. Because of its macroreticular gel structure, agarose is suitable for fractionating molecules or complexes of very large molecular weight (>500 kDa), e.g. virus particles. Also because of their macroreticular structure, agarose gels have exceptional mechanical strength and are resistant to compaction. The pore size of agarose gels can be varied by varying the gel concentration.

A cross-linked form of agarose, with markedly increased thermal and chemical stability, has been described[12] and is commercially available. Although the chemistry involved in the cross-linking reactions is different, the final cross-link structure is the same as that in epichlorhydrin cross-linked Sephadex (see *Figure 75*).

*Table 4* Some commercially available media for MEC

| Gel | Polymer | Cross-linking | Fractionation range (kDa) |
|---|---|---|---|
| **Sephadex** | dextran | epichlorhydrin | |
| G-25 | | | 1→5 |
| G-50 | | | 1.5→30 |
| G-75 | | | 3→70 |
| G-100 | | | 4→150 |
| G-150 | | | 5→300 |
| G-200 | | | 5→500 |
| **PDX** | dextran | | |
| G.F. 25 | | | 1→5 |
| G.F. 50 | | | 1.5→30 |
| **Bio-Gel** | polyacrylamide | bis-acrylamide | |
| P-2 | | | 0.1→1.8 |
| P-4 | | | 0.8→4 |
| P-6 | | | 1→6 |
| P-10 | | | 1.5→20 |
| P-30 | | | 2.5→40 |
| P-60 | | | 3→60 |
| P-100 | | | 5→100 |
| **Sephacryl HR** | dextran | bis-acrylamide | |
| S-100 | | | 1→100 |
| S-200 | | | 5→250 |
| S-300 | | | 10→1,500 |
| S-400 | | | 20→8,000 |
| S-500 | | | 40→20,000 |
| **Trisacryl Plus** | N-tris[hydroxymethyl] methyl methacrylamide | | |
| GF2-M | | | 1→15 |
| GF4-M | | | 5→25 |

The mechanical properties of agarose make it especially suitable as a medium for chromatography. However, its fractionating range is too high for most purposes. A logical development, then, would be to make a composite of agarose, which forms a macroreticular structure, with another gel which forms a microreticular structure. Such a composite should have the strength of agarose, and the separating properties of the microreticular gel. This concept has been realised in the Superdex range of gels from Pharmacia Biotech. Superdex consists of highly cross-linked agarose beads, to which dextran is covalently bonded. The dextran chains determine the separating properties of the composite, while the agarose provides excellent strength.

Sephacryl High Resolution (HR), a product of Pharmacia Biotech, is a composite gel of a different sort. It consists of allyl dextran (the constituent polymer of Sephadex) cross-linked with N, N'-methylene bisacrylamide (the cross-linking agent of acrylamide gels) to form a gel with high mechanical strength. The porosity of the gel is determined by the concentration of the dextran. Gels having five different porosities are available, denoted Sephacryl S-100 HR → S-500 HR. The "HR" refers to the high resolving power of the resins, which is a consequence of their small and uniform particle size (wet bead diameter = 25→75 μm). An S-1000 resin is available; this is not denoted "HR" as the particles are bigger and more variable (wet bead diameter = 40→105 μm). S-1000 separates in the range 500→100,000 kDa.

Trisacryl Plus, a product of Sepracor Inc., consists of poly-(N-tris [hydroxymethyl] methacrylamide) in bead form with a narrow size distribution (40-80 μm) for high resolution. It is available in two pore sizes, GF2-M, which fractionates in the range 1→15 kDa, and GF4-M, which fractionates in the range 5→25 kDa.

### 5.5.5 An MEC run

The MEC column should be packed and equilibrated with at least one colvol of buffer, but preferably more. The buffer should ideally contain at least 0.3 M NaCl, to minimise ion-exchange effects[13], but it must be borne in mind that increasing salt concentration increases hydrophobic interactions. After equilibration, the upper column flow-adapter is adjusted down onto the gel bed.

For best resolution, the sample should be about 2→5% of the column volume but for simple desalting it can be up to 20%. The sample may be applied, and the column subsequently run, at a flow rate of 2→10 cm $h^{-1}$. Theoretically, provided there are no marked non-sieving effects, the next sample could be applied as soon as $V_t$ has eluted but, ideally, at least one colvol of buffer should be run through the column before application of the next sample.

## 5.6 Hydroxyapatite chromatography

Crystalline calcium hydroxyphosphate, $[Ca_5(PO_4)_3(OH)]_2$, the major component of tooth enamel, is known as hydroxyapatite (or hydroxylapatite)[14,15], and its particular usefulness in protein isolation is that it binds proteins by a unique mechanism, different from MEC and simple ion-exchange, and it can therefore separate proteins which may not be separable by other means[14,15]. Hydroxyapatite forms blade-like crystals and because the protein binds to the surface of the crystals, rather than within a gel lattice, the protein binding capacity is relatively low. For this reason, hydroxyapatite is best suited for use as one of the final steps in a purification.

Blade-shaped crystals are not optimal for chromatography and they tend to be brittle, thus generating "fines" which block the column and limit its life to three or four runs. Several manufacturers have attempted to overcome this by making spherical forms of hydroxyapatite, e.g. macro-prep ceramic hydroxyapatite from Bio-Rad and HA ultragel from Pharmacia. However, if the hydroxyapatite is used at the end of a purification, the fact that the classical crystals have a limited life is of lesser consequence.

### 5.6.1 The mechanism of hydroxyapatite chromatography

The separating mechanism of hydroxyapatite is summarised in the review by Gorbunoff[14]. Hydroxyapatite crystals have positive surface charges, due to their constituent calcium ions, and negative charges due to their phosphate groups. The nett charge can be varied by the buffer - it is negative in phosphate buffer, neutral in NaCl and positive in $CaCl_2$ or $MgCl_2$.

Positive amino groups of proteins bind electrostatically to negative charges on the hydroxyapatite, and are thus influenced by its nett charge.

$$\text{Hydroxyapatite - } PO_4^{-} \cdots\cdots H_3^{+}N \text{ - Protein}$$

Negative carboxyl groups, on the other hand, bind by complexing with calcium in the hydroxyapatite.

$$\text{Hydroxyapatite - Ca - OOC - Protein}$$

The retention of acidic (negatively charged) proteins is thus affected by the nett charge on the hydroxyapatite in a manner opposite to that of basic (positively charged) proteins. $CaCl_2$ and $MgCl_2$ increase the binding of acidic proteins by formation of salt bridges between protein carboxyl groups and hydroxyapatite phosphate sites.

$$\text{Hydroxyapatite - }PO_4^- \text{.....}Ca^{2+}\text{.....}^-OOC\text{ - Protein}$$

Basic proteins may be eluted from hydroxyapatite by negative ions such as $F^-$, $Cl^-$, and $HPO_4^{2-}$, which compete with its negative phosphate sites, or by $Ca^{2+}$ or $Mg^{2+}$ ions, which specifically complex with its phosphate sites and neutralise their charges. Acidic (negative) proteins may be eluted by displacement of their carboxyl groups from hydroxyapatite complexing sites by ions, such as phosphate or $F^-$, which form stronger complexes with calcium.

Most proteins contain both amino and carboxyl groups and it will be noticed that phosphate is effective in eluting both types. Consequently, a common means of eluting proteins from hydroxyapatite is by the application of a phosphate gradient - often K-phosphate, because Na-phosphate has a limited solubility at low temperature. Gorbunoff[13] discusses alternative approaches, where the effects of $CaCl_2$ and $MgCl_2$, and of NaCl or KCl, can additionally be exploited in elution schemes. As previously mentioned, using these devices, separations may be achieved which are not possible using other chromatographic systems and hydroxyapatite is thus a valuable technique in the biochemist's portfolio.

## 5.7 Affinity chromatography

The chromatographic methods discussed above are all dependent upon the gross physicochemical properties of the protein. However, the biological activity of the protein is generally more subtle and depends upon the very specific, complementary, steric relationship between the active site and a substrate (or inhibitor), or a binding site and a ligand, as the case may be. Affinity chromatography[16,17] exploits this biospecific relationship between a protein and a ligand, to specifically select out a desired protein from a crude mixture, essentially in a single step.

The specific ligand, which in the case of an enzyme may be a substrate or an inhibitor, is immobilised by conjugation to an insoluble matrix, in a manner which does not interfere with its interaction with the protein. This may require the use of a spacer arm, which typically consists of a chain of about 6-10 carbon atoms. An affinity chromatography resin is thus comprised of three parts, i) the matrix, which is similar to the matrices used for ion-exchange chromatography, ii) a spacer arm, and iii) the ligand. Matrix/spacer arm combinations are commercially available, since these are universal reagents, and simply require the addition of an appropriate ligand.

The sample solution is passed through theMWolumn and by its specific interaction with the immobilised ligand the protein of interest is retained, while all other proteins pass straight through the column. Subsequently, the protein can be eluted by a change in either the pH or ionic strength of the buffer or by addition of a free competing ligand, or of a chaotrope.

Because the protein is immobilised in a small volume of resin, affinity columns are generally quite small. Also, the volume of solution in which the protein occurs may be large and to pass this volume through the column in a reasonable time, while maintaining the linear flow rate within limits, the column is usually relatively wide (e.g. 15 x 15 mm i.d.). To overcome the problem of excessive volumes, there may be some advantage in preceding affinity chromatography by a quick concentrating method, such as TPP.

Affinity chromatography is commonly used to harvest genetically-engineered, expressed proteins. The protein of interest is expressed as a fusion protein with another protein for which a readily-immobilisable ligand is available. Typically, a proteinase-sensitive site is introduced at the junction between the two proteins. After affinity purification of the fusion protein, this may be cleaved by a proteinase to release the expressed protein of interest. Several systems are commercially available from biotech. suppliers, one example being the GST system provided by Amersham Biosciences. Another common practice is to add a tail of about six histidine residues (a "His-tag") to the end of the expressed protein, for later affinity purification on nickel chelating columns. An advantage of the His-tag system is that denatured proteins may be isolated.

Affinity chromatography can be implemented in a great variety of ways - too many to list here - and a useful way of exploring these is to consult the catalogues of various biotech. suppliers.

## 5.8 Hydrophobic interaction (HI) chromatography

HI-chromatography[18] was discovered serendipitously when, in control experiments, ligands were omitted from the matrix/spacer arm combination. It was found that the resulting resins were nevertheless effective at separating proteins, due to hydrophobic interactions between the sample proteins and the aliphatic spacer arms. Following this discovery, HI-resins were purposefully designed to optimise the hydrophobic interaction.

Hydrophobic "bonds" are increased in strength by an increase in buffer ionic strength. HI-chromatography therefore conveniently fits into an isolation scheme, immediately after a salting out step, as the high salt levels will promote binding to the HI-resin. Proteins can subsequently be eluted by decreasing the buffer ionic strength, either in a stepwise manner or in a gradient.

## 5.9 HPLC

HPLC, commonly understood to mean "high pressure liquid chromatography" or "high performance liquid chromatography", has two distinguishing characteristics. Firstly, the stationary phase particle size is very small, so that the total surface area (wetted area) of the particles is very large and the resistance to flow is consequently high, which necessitates the use of high-pressure pumps and columns, packing materials and monitors etc. able to withstand the high pressure. Such equipment is relatively expensive, which has induced some wits to describe HPLC as "high *price* liquid chromatography".

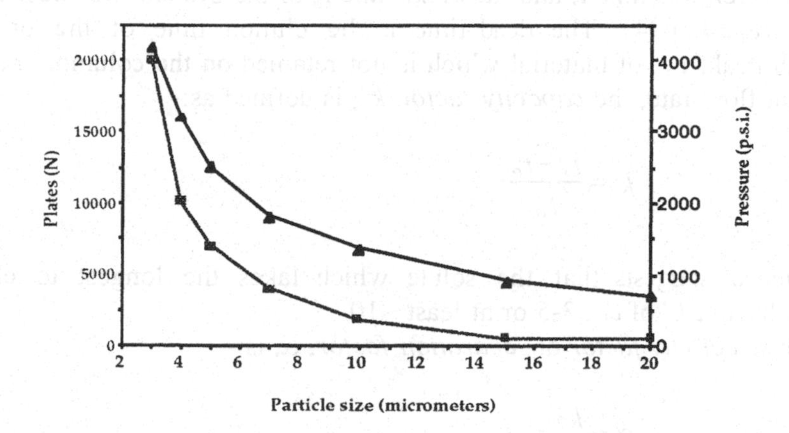


*Figure 76.* The effect of stationary particle size upon the number of theoretical plates ($N$) and the pressure required to run the column optimally.

Balancing this, however, is the chief advantage of HPLC which is *speed*. An HPLC run can be effected in a matter of minutes, compared to the hours normally required for a low pressure run. This is possible because the small size of the stationary phase particles means that equilibration between the stationary and mobile phases is very quick, as reflected in the high $N$ values.

The second advantage of HPLC is *versatility*. Many different types of molecules, both small and large (including proteins) can be separated by HPLC[19-22]. The basis for this versatility is that, in addition to the forms of chromatography discussed above, HPLC uses an additional phenomenon, i.e. the interaction of the sample molecules with the solvent(s). These solute/solvent interactions may occur via five possible mechanisms:-

- van der Waal's interactions
- dipole interactions
- hydrogen bonding
- dielectric interactions
- electrostatic interactions.

The overall affinity between the solute and the solvent is usually a combination of several of these forces and this provides a large scope for selecting a solvent system to optimise any given separation.

### 5.9.1 Concepts and terms relevant to HPLC

The *retention time* $t_r$ and the *dead time* $t_o$ of the system are illustrated in *Figure 54*, p94. The dead time is the elution time of the break-through peak, i.e. of material which is not retained on the column. For a constant flow rate, the *capacity factor*, $k'$, is defined as:-

$$k' = \frac{t_r - t_o}{t_o}$$

Experience suggests that the solute which takes the longest to elute should have a $k'$ of *ca*. 3-5 or at least <10.

The *selectivity factor* or *separation factor*, α, is:-

$$\alpha = \frac{k'_2}{k'_1}$$

and α must be >1 for two peaks to be separated.

*Column efficiency* is given by Eqn 5.1 (p 94), i.e.:-

$$N = a\left(\frac{t_r}{W}\right)^2$$

Examination of this equation reveals that as the run progresses, the peaks get wider and flatter and eventually trace components can become difficult to detect. Gradient elution can overcome this problem, so that all peaks are equally sharp (i.e. $N$ increases as the run progresses). In contrast to a gradient, an *isocratic* solvent is one of constant composition, which does not change during the run.

The resolution of peaks is described by Eqn 5.2 (p94), i.e.:-

$$R = \frac{t_{r2} - t_{r1}}{\left(\frac{W_2 + W_1}{2}\right)}$$

However, this equation does not reveal how the separation may be improved. By combining column efficiency $N$, selectivity $\alpha$, and the capacity factor $k'$ a more useful expression, known as *the resolution equation*, can be obtained, i.e.:-

$$R = \tfrac{1}{4}(\sqrt{N}).(\frac{\alpha - 1}{\alpha}).(\frac{k_2'}{k_2' + 1}) \quad 5.14$$

## 5.9.2 Stationary phase materials

HPLC columns are discussed by Neue[23]. There are relatively few stationary phases available for HPLC and, of these, *reverse phase* packings are the most commonly used. These consist of silica particles in which the exposed SiOH groups are derivatised with aliphatic groups ($C_2$, $C_4$, phenyl, $C_8$ or $C_{18}$). In some packing materials residual SiOH groups that have not been derivatised are "endcapped" by reacting them with a small silanising reagent such as $C_1$ or $C_2$.

Different reverse phase packings may differ in the size and/or shape of the silica particles, their pore size, their carbon load (the amount of carbon immobilised per gram of silica) and the presence or absence of endcapping. All of these factors affect the separation that may be achieved. Reverse phase packings are most retentive of non-polar solutes and are thus comparable to HI resins.

Normal phase packings, by contrast, consist of active silica (SiOH) itself which is most effective in the separation of relatively polar compounds. Silica columns, however, are "poisoned" by exposure to even small amounts of water. "Capping", with a cyano or amino group, to form so-called "bonded phase" materials, makes them more water-resistant and enables the use of more non-polar solvents.

Packing materials for ion-exchange and size exclusion chromatography (SEC) (essentially the same as molecular exclusion chromatography, or MEC) are also available for use in HPLC and may be used for the separation of proteins.

*Pellicular* resins consist of a strong core particle covered with a layer of the separating material. Diffusion to and within this layer is thus facilitated.

### 5.9.3 Solvent systems

Experience has shown that only four solvents are adequate for the vast majority of HPLC separations, namely, methanol, acetonitrile, tetrahydrofuran and water. The properties of these solvents are listed in Table 5.

*Table 5* Properties of common HPLC solvents

| Solvent | UV cut-off (nm) | BP (°C) | Dielectric constant | viscosity | $P'$ |
|---|---|---|---|---|---|
| Acetonitrile | 190 | 82 | 37.5 | 0.34 | 5.8 |
| Methanol | 205 | 65 | 32.7 | 0.54 | 5.1 |
| Tetrahydrofuran | 212 | 66 | 7.6 | 0.46 | 4.0 |
| Water | 190 | 100 | 80.0 | 0.89 | 10.2 |

The *polarity factor*, $P'$, can be used to modify the $k'$ values of peaks. Notice that water has about twice the polarity of the other solvents and so water is commonly added to modify $k'$ values. The relationship between the P' values of two solvents and the resulting difference in the k' value of a solute is given by the equation:-

$$\frac{k_2'}{k_1'} = 10^{(P_2' - P_1')/2}$$

so a 2-fold change in $P'$ results in about a 10-fold change in $k'$. Addition of water (i.e. making the solvent "weaker") generally increases $k'$ values.

Re-considering Eqn 5.14, i.e.:-

$$R = \frac{1}{4}(\sqrt{N}).(\frac{\alpha - 1}{\alpha}).(\frac{k_2'}{k_2' + 1})$$

it is evident that resolution is affected by three factors, the column efficiency, $N$, the capacity factor, $k'$, and the selectivity, $\alpha$, and to improve resolution any or all of these factors could be changed.

$N$ may be improved by optimising the flow rate, increasing the column length or decreasing the particle size of the packing material. Increasing $k'$ will increase resolution but at the cost of an increased run time and decrease in peak height. Increasing the selectivity, $\alpha$, increases resolution without significantly altering the run time or the peak height and is likely to be the most rewarding approach. Selectivity may be changed by altering the properties of the stationary or mobile phases. With a given stationary phase, the only scope is to change the properties of the mobile phase. This can be done in a systematic manner as described by Glajch *et al.*[24], Lehrer[25] and Snyder and Glajch[26]. Manufacturers' catalogues should also be consulted as these provide large databases of separations.

For the separation of proteins and peptides, experience suggests that greatest success is likely to be achieved with a large pore packing material (typically 300Å), usually C8 - since most proteins have considerable surface hydrophobicity and a C18 stationary phase is too "sticky". The solvent is usually a gradient of acetonitrile in water, with the addition of about 0.1% trifluoroacetic acid to protonate the functional groups.

### 5.9.4 Preparative HPLC

HPLC is most commonly used as an analytical method but it can be scaled up to prepare milligram or even gram amounts of protein. The best approach is to optimise the separation at an analytical scale (a typical analytical column might be 4.6. x 150 mm, run at 1 ml min$^{-1}$) and to scale up to a preparative column (typically 25 x 250 mm or larger), keeping the linear flow rate (see p99) constant.

*References*

1. Kyte, J. (1995) in ***Structure in Protein Chemistry***. Garland Publishing Inc., New York and London, pp2-11.
2. Van Deemter, J. J., Zuiderweg, F. J. and Klingenberg, A. (1956) Longitudinal diffusion and resistance to mass transfer as causes of non-ideality in chromatography. Chem. Engng. Sci. 5, 271-289.
3. Sluyterman, L. A. A and Elgersma, O. (1978) Chromatofocusing: isoelectric focusing on ion-exchange columns. I. General principles. J. Chromatogr. 150, 17-30.
4. Sluyterman, L. A. A and Elgersma, O. (1978) Chromatofocusing: isoelectric focusing on ion-exchange columns. II. Experimental verification. J. Chromatogr. 150, 31-44.

5. Sluyterman, L. A. A and Widjdenes, J. (1981) Chromatofocusing. III. The properties of a DEAE-agarose anion exchanger and its suitability for protein separations. J. Chromatogr. 206, 429-440.
6. Sluyterman, L. A. A and Widjdenes, J. (1981) Chromatofocusing. IV. The properties of an agarose polyethyleneimine ion exchanger and its suitability for protein separations. J. Chromatogr. 206, 441-447.
7. Polson, A., and Katz, W. (1969) "Tanned" gelatin spheres and granules for exclusion chromatography. Biochem. J. 108, 641-646.
8. Laurent, T. C. and Killander, J. (1964) A theory of gel filtration and its experimental verification. J. Chromatog. 14, 317-330.
9. Hjertén, S. (1970) Thermodynamic treatment of partition experiments with special reference to molecular sieve chromatography. J. Chromatogr. 50, 189-208.
10. Andrews, P. (1970) Estimation of molecular size and molecular weights of biological compounds by gel filtration. Methods of Biochemical Analysis 18, 1-53.
11. Fischer, L. (1969) An introduction to gel chromatography. in *Laboratory techniques in Biochemistry and Molecular Biology, Vol 1.* (Work, T. S. and Work, E. eds) pp151-396.
12. Porath, J., Janson, J.-C. and Låås, T. (1971) Agar derivatives for chromatography, electrophoresis, and gel-bound enzymes. 1. Desulphated and reduced crosslinked agar and agarose in spherical bead form. J. Chromatogr. 60, 167-177.
13. Herold, M. (1993) SEC: influence of salt concentration in the mobile phase. Int. Laboratory, March 1993, 34-36.
14. Gorbunoff, M. J. (1985) Protein chromatography on hydroxyapatite columns. Methods Enzymol. 117, 370-380.
15. Bernardi, G. (1973) Chromatography of proteins on hydroxyapatite. Methods Enzymol. 27, 471-479.
16. Cuatrecasas, P. and Anfinsen, C. B. (1971) Affinity chromatography. Annu. Rev. Biochem. 40, 259-278.
17. Jakoby, W. B. and Wilchek, M, (eds.) Methods Enzymol. 34.
18. Ochoa, J. L. (1978) Hydrophobic (interaction) chromatography. Biochimie 60, 1-15.
19. Gertz, C. (1990) *HPLC tips and tricks.* Alden, New York
20. Krstulovic, A.M. and Brown, P.R. (1982) *Reversed-phase high-performance liquid chromatography: theory, practice and biomedical applications.* Wiley, New york.
21. Snyder, L.R. and Kirkland, J.J. (1979) *Introduction to modern liquid chromatography, 2nd Ed.* Wiley, New York.
22. Snyder, L.R., Glajch, J.L. and Kirkland, J.J. (1988) *Practical HPLC method development.* Wiley, New York.
23. Neue, U.D. (1997) *HPLC columns. Theory, technology and practice.* Wiley-VCH Inc., New York.
24. Glajch, J.L., Kirkland, J.J., Squire, K.M. and Minor, J.M. (1980) Optimisation of solvent strength and selectivity for reversed-phase liquid chromatography. J. Chromatogr. 199, 57-79.
25. Lehrer, R. (1981) The practice of high performance LC with four solvents. American laboratory Oct. 1981, 113-125.
26. Snyder, L.R. and Glajch, J.L. (1981) Solvent strength of multicomponent mobile phases in liquid-solid chromatography. Binary-solvent mixtures and solvent localization. J. Chromatogr. 214, 1-19.

## 5.10 Chapter 5 study questions

1. Define the term "distribution coefficient" as applicable to chromatography.
2. Define "HETP". What value of HETP is best? What factors influence HETP?
3. What is the optimum value for each of the factors that affect the HETP?
4. The distribution coefficient of substance A is 0.4 and of substance B is 0.6. Which will move more slowly through a chromatography column?
5. Why is it necessary to have a minimum dead volume on the outlet side of a chromatography column?
6. What are some desirable properties of the matrix of an ion-exchanger?
7. Is DEAE generally suitable for binding a protein, a) below the pI of the protein, b) at its pI, or, c) above its pI?
8. Two proteins have pI values of 7.6 and 8.1. Briefly describe an ion-exchange procedure that may be used to separate these.
9. With regard to the substituent on a cation exchanger, is a strong acidic or basic group generally better or worse than a weak group? Explain.
10. Why, after eluting an ion-exchange column with a gradient from a gradient mixer, it is necessary to elute with a further column volume of finishing buffer?
11. For eluting an ion-exchange column, is an ionic strength gradient usually better/worse than a pH gradient? Explain.
12. A chromatographic column (25 mm i.d. x 95 cm) is run at a volumetric flow rate of 50 ml $h^{-1}$. (a) At what volumetric flow rate should an 18 mm i.d. column be run to give equivalent chromatographic conditions? (b) If the sample volume applied to the 25 mm column was 30 ml, what volume of sample should be applied to the 18 mm column? (c) Assuming the 2.5 x 95 cm column was filled with a molecular exclusion gel, what time period should be set on an automatic shut-off timer to ensure that all peaks

would be completely eluted? (d) If a fraction collector with 90 tubes was available, what time per tube should be set to collect the entire run? (e) What would be the volume in each tube? (f) How long after application of the sample would one expect to see the peak elution of a sample component which is larger than the exclusion limit of the gel? (g) How long after application of the sample would one expect to see elution of a peak having a $K_{av}$ of 0.5?

13. Describe the difference between a microreticular and a macroreticular gel and say which of the following gels is which:- Sephadex, agarose, polyacrylamide.

14. A molecular exclusion column was standardised with the following standard globular proteins:-

| Protein | *MW* (kDa) | $K_{av}$ |
|---|---|---|
| Cytochrome C | 13 | 0.62 |
| Myoglobin | 17 | 0.55 |
| Chymotrypsinogen | 25 | 0.45 |
| Ovalbumin | 45 | 0.32 |
| BSA | 67 | 0.20 |

Determine the *MW* of a protein having a $K_{av}$ of 0.40.
It is known that this unknown protein consists of two subunits of equal size, which dissociate in 8 M urea. If the column size was 1.5 x 50 cm, in what volume would you expect the unknown protein to elute in a buffer containing 8 M urea?

17. A molecule has a $K_{av}$ of 0,6. What % of the stationary phase is available to it?

18. The calculated $K_{av}$ value of a glucosidase enzyme on Sephadex G-100 was >1. What does this tell you? Can you offer a possible explanation for this phenomenon?

19. When in a protein isolation would the use of HI chromatography be most appropriate?

20. Assuming evenly sized spherical gel particles, how is $V_o$ affected by the resin particle size?

21. Calculate the phase ratio for a molecular exclusion gel. (Clue: remember that $V_o = 0.36\ V_t$).

22. Calculate the partition ratio for a solute having a $K_{av}$ of 0.75 on an MEC gel.

# Chapter 6

# Electrophoresis

Active fractions isolated by a preparative fractionation procedure may be subjected to a number of analytical fractionation procedures to determine their purity. Analytical fractionations are distinguished from preparative fractionations by the criteria shown in Table 5.

*Table 5.* The difference between preparative and analytical fractionations.

| | Analytical | Preparative |
|---|---|---|
| Scale | Small | Large |
| Fate of sample | Destroyed | Preserved |
| Product | Information | Active fraction |

In an analytical fractionation, therefore, a small amount of sample is sacrificed in order to gain information about the state of purity of the material being analysed. Of the many physico-chemical techniques which have contributed to our knowledge of proteins (and nucleic acids), electrophoretic techniques occupy a position of primary importance. Electrophoresis finds its greatest usefulness in the analysis of mixtures and in the determination of purity, although certain forms of electrophoresis may be applied on a preparative scale.

## 6.1 Principles of electrophoresis

In metal conductors, electric current is carried by the movement of electrons, largely along the surface of the metal because of the mutual repulsion of electrons. In solutions, the electric current flows between electrodes and is carried by ions. Electrophoresis concerns the movement of such ions in electrolytic cells (as distinct from electrochemical cells). In electrolytic cells, the negative electrode - **the cathode** - donates electrons and the positive electrode - **the anode** - takes up electrons to complete the circuit. The ions that result from the acceptance of electrons from the cathode will be negatively charged and will thus migrate towards the positive anode. Because of their anodic migration,

| "PANIC" - | Positive Anode |
|---|---|
| | Negative Cathode |
| (NI) | Not the Ions |

negative ions are called "anions". Ions which result from the donation of an electron to the electron-deficient (i.e. positively charged) anode will themselves be electron deficient, and thus positively charged. These will migrate to the cathode and are thus called cations.

Students are sometimes confused by the different labelling of the anode and cathode in electrochemical cells *vs* electrolytic cells. The distinction between the two types of cell is summarised in Table 6. Electrophoresis concerns electrolytic cells, but the interest lies in the movement of the ions, rather than in the electrode reactions - the subject of interest in electrochemistry.

*Table* 6. Similarities and differences between electrochemical and electrolytic cells.

| Electrochemical cells | Electrolytic cells |
|---|---|
| Produce electricity from a chemical reaction | Use electricity to produce a chemical reaction |
| The reaction is spontaneous | Electrical energy must be supplied for the reaction to occur |
| Oxidation (the loss of electrons) occurs at the anode | Oxidation (the loss of electrons) occurs at the anode |
| Reduction (the gain of electrons) occurs at the cathode | Reduction (the gain of electrons) occurs at the cathode |
| The anode is labelled negative, because it supplies electrons to the external circuit | The cathode is labelled negative as it is connected to the negative terminal of the applied power source |
| The cathode is labelled positive because electrons move towards it in the external circuit | The anode is labelled positive as it is connected to the positive terminal of the applied power source |
| In the electrolyte, the current is carried by anions moving to the anode and cations moving to the cathode | In the electrolyte, the current is carried by anions moving to the anode and cations moving to the cathode |

from Garnett and Garnett[1]

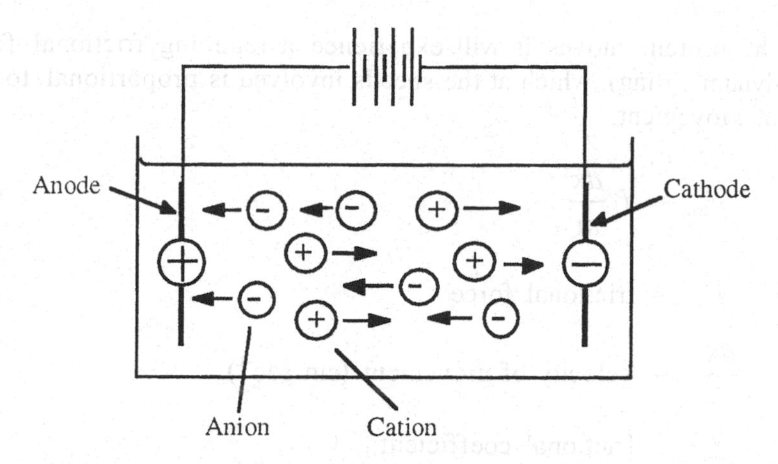


*Figure 77.* Electrophoresis: the movement of ions in an applied electric field.

In electrophoresis, a potential difference (voltage) is applied between the anode and the cathode and if the solution between these is of constant composition and constant cross-section (i.e. constant resistance), the voltage gradient between them ($dV/dx$) will be linear, with units of volts $cm^{-1}$. (In Section 6.12 the effects of non-linear voltage gradients will be explored.)

An ion placed in such an electric field will experience a force:-

$$F = K.q\frac{dV}{dx} \qquad 6.1$$

Where,
$F$ = electrophoretic force
$K$ = a constant (embodying the Faraday constant and Avogadro's number)
$q$ = nett charge on the protein (atomic charges/protein molecule)
$\frac{dV}{dx}$ = the voltage gradient (volts $cm^{-1}$).

This force will cause the protein to accelerate towards either the cathode or the anode, depending on the sign of its charge.

As the protein moves it will experience a retarding frictional force (hydrodynamic drag), which at the speeds involved is proportional to the speed of movement.

$$F_{fric} = f.\frac{dx}{dt} \qquad 6.2$$

Where, $F_{fric}$ = frictional force

$\frac{dx}{dt}$ = velocity of movement (cm sec$^{-1}$)

$f$ = frictional coefficient

It will be recalled that this situation is very similar to that obtaining during centrifugation (Section 3.6.2), and the frictional coefficient can be determined in the same way,

$$D = \frac{RT}{f}$$

Hence, $$f = \frac{RT}{D}$$

The proteins very soon reach terminal velocity, at which point the electrophoretic (propelling) force equals the frictional (retarding) force, i.e. from equations 6.1 and 6.2:

$$K.q\frac{dV}{dx} = f.\frac{dx}{dt} \qquad 6.3$$

The free electrophoretic mobility, ($\mu$), with units of (cm$^2$ volt$^{-1}$ sec$^{-1}$) can be defined as the velocity per unit of voltage gradient, i.e.:

$$\mu = \text{velocity (voltage gradient)}^{-1}$$

$$= \frac{dx/dt}{dV/dx}$$

Hence, from equation 6.3,

$$\mu = K\frac{q}{f} \qquad 6.4$$

The electrophoretic mobility is thus a function of the charge on the protein ion and the medium through which it is travelling. Electrophoretic techniques exploit the fact that different ions have different mobilities in an electric field and so can be separated by electrophoresis.

The flow of electricity in electrophoresis is subject to the same physical laws as other forms of electricity. For example, Ohm's law applies:-

$$I = \frac{V}{R} \qquad 6.5$$

Where $I$ = current (amps)
$V$ = potential difference (volts)
$R$ = resistance (ohms).

The unit of electrical charge is the coulomb and the unit of current [the ampere (amp)] may be defined as coulombs sec$^{-1}$, i.e.,

$$I = \frac{\text{coulombs}}{t} \qquad 6.6$$

$\therefore$ coulombs $= I.t$

The flow of electricity involves work, which generates heat, and the work (W, in joules) done in transferring a charge of $q$ coulombs between a potential difference of V volts is:-

$$W = qV = (I.t)V = IVt \qquad 6.7$$

And, since, from Eqn 6.5,

$$V = IR,$$

Then, $IVt = I(IR)t = I^2Rt$

∴ from Eqn 6.7,

$$W = I^2Rt \qquad \text{(Joule's law of heating)}$$

Which means that $I^2Rt$ joules of heat will be developed in the conductor.

The power (in watts) (defined as the rate of work) gives the rate of heating (joules $sec^{-1}$).

$$\text{Thus,} \quad Watts = \frac{I^2 Rt}{t} = I^2R \qquad 6.8$$

$$= \frac{V^2}{R} \text{ (since } I = \frac{V}{R}\text{)}$$

### 6.1.1 The effect of the buffer

The buffer in which electrophoresis is conducted, has a large influence on the migration of proteins. Firstly, the buffer *pH* will influence the charge $q$ on the protein and hence the direction and speed of its migration. Secondly, the buffer *ionic strength* influences the proportion of the current carried by the proteins - at low ionic strength the proteins will carry a relatively large proportion of the current and so will have a relatively fast migration. At high ionic strength, most of the current will be carried by the buffer ions and so the proteins will migrate relatively slowly.

An analogy might be useful in visualising this effect of ionic strength. Imagine a bank where there are two counters - one for deposits (≡ the anode) and one for withdrawals (≡ the cathode), with electrons being the money. The ions may be considered as customers waiting to be served at either counter, which one can visualise as being at opposite ends of the banking hall.

In *Figure 78*, the circles represent customers queuing for service. In electrophoresis, these queues would be along the so-called *field lines*, which are usually (but not necessarily) straight lines. The lighter coloured circles represent buffer ion "customers" and the dark circles represent protein "customers". When the "customer" at the counter is served, they move away, creating a "hole". This "hole" is filled by the next customer in line, and so on, and so the "hole" moves backwards along the line. No matter how far away from a counter any customer is, they will be drawn towards the counter by the periodic appearance of a "hole" in the queue, immediately in front of them. If the counter assistants were very energetic (giving a high current) these "holes" would appear frequently and the customers would all progress quickly. On the other hand, if the counter assistants were lethargic (giving a low current) the "holes" would appear infrequently and progress of the customers would be slow.

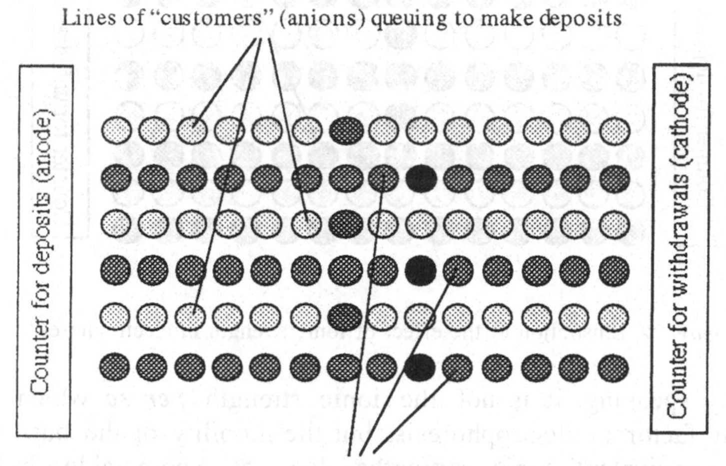


*Figure 78.* A banking hall analogy of electrophoresis.

In *Figure 78*, the relative proportions of protein ions to buffer ions shown is such that there is one protein ion in each queue. However, if we have the counter assistants working at the same rate (i.e. with the same current) but increase the number of customers (i.e. increase the ionic strength), then we will get the situation shown in *Figure 79*.

With more buffer ions present, they will get most of the service (carry most of the current) and the progress of all ions in their respective queues, including the protein ions, will be slower.

In electrophoresis, therefore, a low ionic strength is preferred as it increases the rate of migration of proteins. A low ionic strength is also preferred as it gives a lower heat generation. Assuming a constant voltage, if the ionic strength is increased, the electrical resistance decreases but the current will increase. According to Eqn 6.8, heating is proportional to $I^2$, but is only linearly affected by changes in resistance.

A high ionic strength buffer will therefore lead to greater heat generation, and so a low ionic strength is preferred.

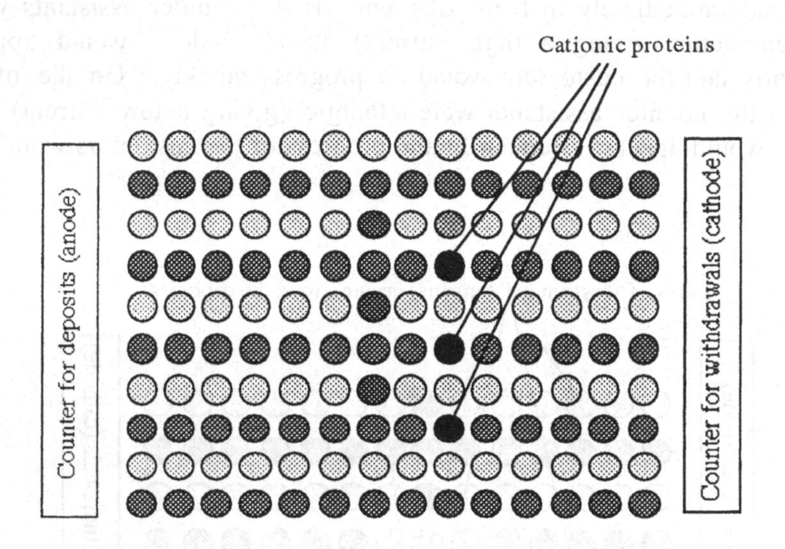


*Figure 79.* Illustration of the effect of ionic strength in electrophoresis.

Strictly speaking, it is not the ionic strength *per se* which is the important factor in electrophoresis, but the mobility of the buffer ions. Thus at equivalent ionic strengths (i.e. at comparable buffering capacities), large buffer ions will migrate more slowly than small buffer ions (because of their greater frictional coefficient, $f$). Large buffer ions will thus lead to less heat generation and a faster migration of the proteins. For example, barbitone (**I**) has a mobility about one quarter of that of Tris (**II**), and can therefore be used at four times the concentration of Tris - at which concentration it will be roughly four times more effective as a buffer.

O
$C_2H_5$ N
=O
$C_2H_5$ N
O

**I**

$CH_2OH$
$CH_2OH-C-NH_2$
$CH_2OH$

**II**

The effect of ionic strength is actually more complex than indicated in the simplistic model given above. The reason is that ionic strength also has an effect on the electrical double layer which surrounds proteins in solution. The ions in the electrical double layer have the effect of decreasing the apparent charge on the protein. As the protein moves under electrophoresis, it takes with it a part of the electrical double layer. As the ionic strength increases, the thickness of the electrical double layer decreases and more of the counterions are drawn along with the migrating protein, effectively reducing its charge. The mobility of the protein thus decreases with increasing ionic strength. A more complete discussion of this effect is given by Kyte[2].

## 6.2 Boundary (Tiselius) electrophoresis

This method, which is now of historic interest only, is mentioned here because it has a bearing on the terminology used to describe electrophoresis. Boundary electrophoresis is an interesting example of free electrophoresis, i.e. where there is no supporting medium, and it is conceptually different from all modern forms of electrophoresis which are all so-called "zone electrophoresis" methods, in which zones of different proteins become completely separated from one another.

In boundary electrophoresis, a protein mixture was introduced into a U-tube and subjected to an electric field (*Figure 80*). The proteins are not completely separated, but, theoretically, the number of proteins in a mixture can be determined by analysis of the number of boundaries formed after a period of electrophoresis. The boundaries can be detected by schlieren optics, which detects changes in refractive index.

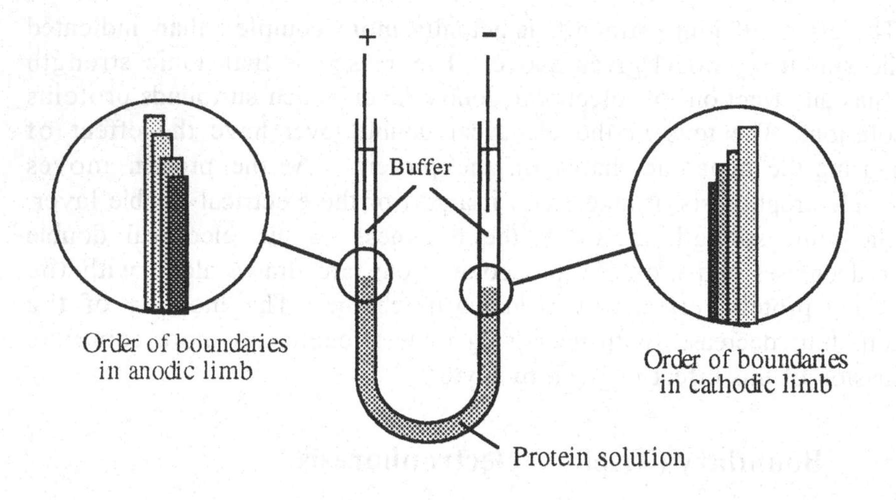


*Figure 80.* Moving boundary (Tiselius) electrophoresis.

A modern form of electrophoresis, which uses a somewhat similar apparatus, but which effects complete separation of protein zones, is capillary electrophoresis, which will be discussed below.

## 6.3 Paper electrophoresis

One of the earliest forms of zone electrophoresis for the separation of proteins was paper electrophoresis. In this a strip of filter paper was used as a medium to support a thin layer of buffer. Since the paper served only to support the buffer, paper electrophoresis can be considered as a form of free electrophoresis (as opposed to electrophoresis in a sieving gel, which will be discussed in Section 6.6) The experimental set-up for paper electrophoresis is shown in *Figure 81*.

A strip of filter paper, typically 20 x 150 mm, is marked with pencil to indicate the anodic and cathodic ends and a line is lightly drawn transversely in the middle, where the sample is to be applied. The strip is soaked in buffer, blotted briefly and suspended between supports in the apparatus. Buffer is added to both the anode and the cathode compartments: it is important that the levels in the two compartments are the same to prevent siphoning through the filter paper. The filter paper is connected to the buffer by filter paper wicks, which must be the same width as the filter paper strip, but can be made of several layers of filter paper.

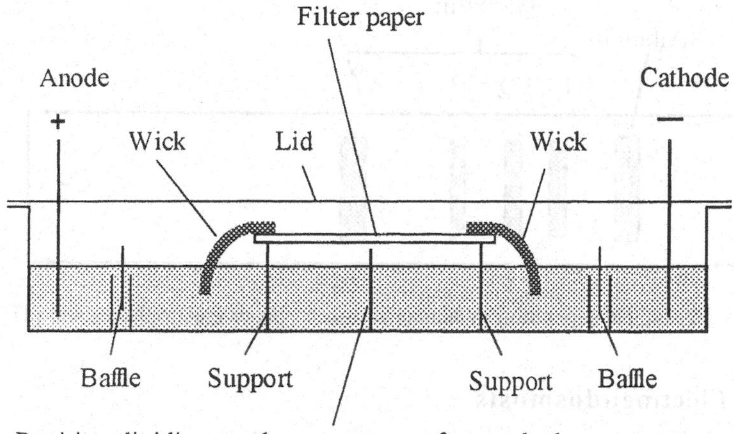


*Figure 81.* Diagrammatic cross-section of an apparatus for paper electrophoresis.

Sample can be applied as a thin line across the middle of the paper strip, but not within *ca*. 5 mm of the edges. There are different ways of applying the sample: a simple way, but which requires some manual dexterity, is to use a Pasteur pipette, drawn down to a thin capillary. After application of the sample, the apparatus is sealed with a lid. This enables the atmosphere within the apparatus to become saturated with water vapour, thereby preventing evaporation of water from the buffer on the strip. The buffers have a maximal exposed surface area to encourage rapid equilibration of the water vapour. As a safety precaution, the apparatus lid is usually coupled with the electrode connections, so that removal of the lid breaks the electric circuit. Without this precaution, fatal shocks might result from inadvertent contact with the electrode solutions. Electrophoresis is run, usually for a number of hours, typically using a voltage gradient of *ca.* 10 volts $cm^{-1}$. To keep electrolysis products away from the protein samples being separated, the buffers in the electrode compartments are separated from the wicks by a baffle system.

After separation, the protein bands are fixed in position and stained with a protein-specific stain, such as Ponceau S or Amido Black, and destained. Paper electrophoresis was used in medical diagnostics and a typical result for the separation of serum is shown in *Figure 82*.

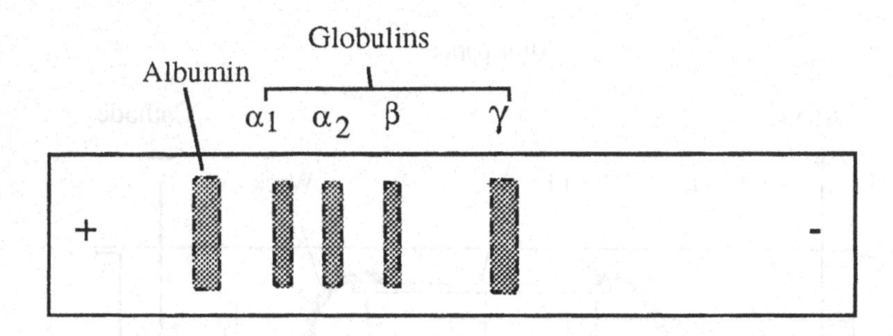


*Figure 82.* A typical separation of human serum by paper electrophoresis at pH 7.2.

### 6.3.1 Electroendosmosis

Paper is comprised largely of cellulose, a β1→4 linked polymer of glucose. Glucose is hydrophilic due to the polarity of its many -OH groups, which readily form hydrogen bonds with water. Cellulose as a whole is not water soluble, however, because of its extensive interchain hydrogen bonds. In forming hydrogen bonds with water, the H of the -OH groups of glucose is shared with the oxygen of water, giving the water a δ+ charge and the cellulose oxygen a δ- charge.

$$\text{Cellulose - O - H} \cdots\cdots \text{O}\begin{smallmatrix}\text{H}\\ \text{H}\end{smallmatrix} \rightleftharpoons \text{Cellulose - O}^{-} \quad \text{H}-\overset{+}{\text{O}}\begin{smallmatrix}\text{H}\\ \text{H}\end{smallmatrix}$$

Cellulose thus acquires an overall negative charge, called the "zeta potential". When placed in an electrical field, a strip of paper (cellulose) tends to move towards the anode, but cannot do so as it is fixed in place. The hydroxonium ions, $H_3O^+$, however, are free to move towards the cathode and do so, resulting in a nett drift of the buffer towards the cathode. This drift of the buffer towards the cathode, known as electroendosmosis, increases the apparent mobility of cations and decreases that of anions. For example, in *Figure 82*, γ-globulin is seen to have an apparent migration to the cathode. However, the pI of γ-globulin is 6.8, and so at pH 7.2 it would be expected to have a nett negative charge and a consequent anodic migration. In fact it does have a slight anodic migration but the electroendosmotic flow is faster than this, resulting in the apparent cathodic migration.

## 6.4 Cellulose acetate membrane electrophoresis (CAM-E)

Cellulose has a rather open, porous structure and a negative charge which, besides causing marked electroendosmosis, can result in the binding of positively charged proteins. These properties of cellulose result in band spreading and consequent poor resolution of bands.

Cellulose acetate is a finer-grained derivative of cellulose made by the esterification of a proportion of the free -OH groups of cellulose to the acetate ester.

$$\text{Cellulose-OH} + CH_3COOH \rightarrow \text{Cellulose-O-}\overset{\overset{\displaystyle O}{\|}}{C}\text{-}CH_3$$

Cellulose acetate membranes give sharper protein bands and better resolution than paper, bind proteins less and have less endosmosis. As a result, CAM-E replaced paper electrophoresis and remained in use, mainly in medical diagnostic laboratories, for many years. The apparatus required for CAM-E is the same as for paper electrophoresis.

Cellulose acetate membranes are white and opaque but can be readily clarified and rendered transparent by immersion in "liquid paraffin". The clear strips, with bands of stained protein, can be scanned to give an objective and quantitative assessment of the separation obtained (*Figure 83*). A more modern way of achieving the same end would be to capture a digital image of the CAM strip which could then be subjected to appropriate image analysis.

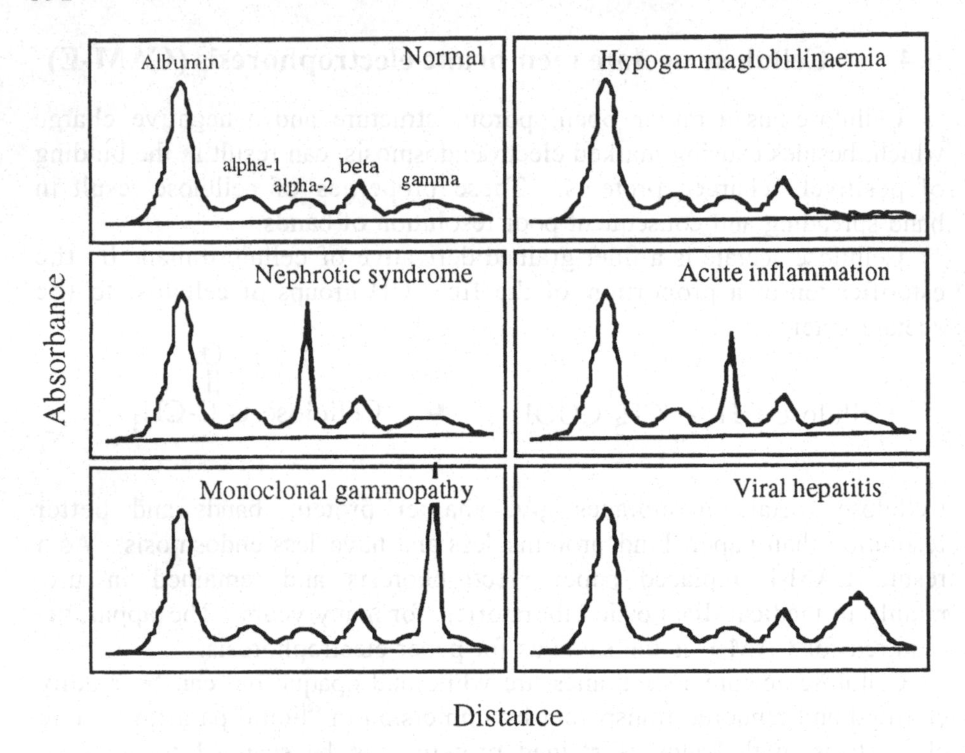


*Figure 83.* Densitometric scans of CAM-E results from different pathological sera.

## 6.5 Agarose gel electrophoresis

CAM, in turn, has largely been replaced by agarose as a support medium for free electrophoresis in medical diagnosis[3]. Although agarose is a gel, it has a macroreticular structure (see *Figure 66*) and thus does not impede the electrophoretic migration of molecules of less than *ca.* 500 kDa. Proteins are thus not usually retarded but nucleic acids can be separated on the basis of their size by gel sieving (see Section 6.6). As with all gels, water cannot flow through agarose. There is therefore no necessity to maintain the two buffer reservoirs at the same height, since the buffer is unable to siphon through an agarose gel.

Agarose has other advantages: it can be obtained in a form with zero electroendosmosis, it can be cast onto a sheet of flexible plastic Gel-Bond® and, after staining and destaining, it can be dried onto the Gel-Bond to provide a durable record which is easily filed.

## 6.6 Starch gel electrophoresis

Historically, starch gel electrophoresis preceded agarose electrophoresis but here the order of discussion is turned about to group mechanistically related techniques. Starch gel electrophoresis was introduced by Smithies[4] in 1955. Starch forms microreticular, thermosetting gels comprised of interlocking starch helices, cross-linked by H-bonds. The microreticular nature of starch gels introduced the phenomenon of *gel sieving* which revolutionised electrophoresis by greatly increasing its resolution and sensitivity.

Starch gel electrophoresis is not merely an historical curiosity as it is still widely used today, not by biochemists but by biologists exploring the taxonomic relationships of organisms or in plant breeding. The reason for its continued use is that starch gels are non-toxic and biodegradable and are thus suitable for large-scale screening.

| An advantage of starch gels is that they are non-toxic and biodegradable and are thus suitable for large-scale screening. |
|---|

In a microreticular gel, a protein migrating under electrophoresis faces a greatly increased frictional resistance, due to the fact that the proteins have to migrate through the 3-D gel network. This resistance is an inverse function of the size of the protein, so that small proteins will migrate with less friction than larger proteins, while proteins larger than the exclusion limit of the gel will not be able to enter into the gel at all.

Ferguson[5] has determined that the mobility of a protein in a starch gel, as a function of the gel concentration, is described by the equation:-

$$u_i = u_i^o e^{-{}^iK_i T_s} \qquad 5.9$$

Where $u_i$ = the electrophoretic mobility of protein $i$ in a starch gel of concentration $T_s$ percent

$u_i^o$ = the free electrophoretic mobility of protein $i$.

${}^iK_i$ = a constant unique to protein $i$.

i.e. the mobility decreases logarithmically as the gel concentration increases. A plot of ln $u_i$ vs $T_s$ gives a straight line, of slope $-{}^iK_i$ known as a Ferguson plot.

The same apparatus as used for paper electrophoresis and CAM-E can be used for starch gel electrophoresis. The gel is cast as a horizontal slab, which is connected to the buffer reservoirs using filter paper wicks. The slab must not be too thick to prevent excessive heat build-up. For extra cooling, the slab may be supported on a block internally cooled by circulating cold water. To accommodate the samples to be separated, small slit-like wells are cast in the gel slab, using a purpose-made mould, or cut with a scalpel. In the latter case the sample can be introduced into the slit by inserting a small strip of filter paper impregnated with sample.

After sample is introduced into the sample wells, the electric field is applied across the length of the gel. Under the influence of the electric field, the sample proteins will move either towards the cathode or the anode, depending upon their charge at the buffer pH. Initially, they will migrate through the buffer in the sample wells by free electrophoresis. When they strike the well wall, on either the anodic or cathodic side, the resistance to their migration will increase and the sample will be concentrated into a narrow band.

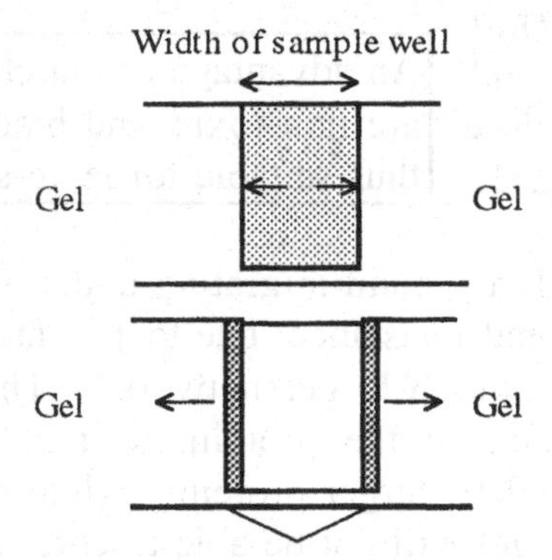


*Figure 84.* Sharpening of starting zones in starch gel electrophoresis.

The resolution (see p93) in electrophoresis is a function of the starting bandwidth and of the band spreading during electrophoresis. The latter is largely due to diffusion of the protein from its zone of highest concentration. The structure of a microreticular gel, however, not only impedes the progress of proteins undergoing electrophoresis, but also limits diffusion. The combination of narrow starting bands and reduced diffusion results in the marked improvements in resolution and sensitivity of gel electrophoresis. The sensitivity is increased since proteins are more easily detected the higher their concentration and, by the initial

> Starch gels improve resolution by forming narrow starting bands and minimising diffusion spreading of bands during electophoresis

concentration of the bands and subsequent minimisation of diffusion, proteins present at low levels can be detected.

A down-side of SGE is that, due to the variability of starch which is a natural product, results tend to vary from lab to lab and in the same lab at different times. This motivated a search for a more uniform, synthetic gel.

## 6.7 Polyacrylamide gel electrophoresis (PAGE)

Polyacrylamide, as a medium for gel electrophoresis was introduced by Ornstein[4]. It has the advantage of being a synthetic gel, which is highly reproducible. Moreover, the pore size can be controlled by varying the proportions of acrylamide and the crosslinking agent, bisacrylamide (see p124). Polyacrylamide can be used as a direct replacement for starch, though it is less conveniently used than starch in the horizontal slab format because polymerisation of polyacrylamide is inhibited by oxygen, so it must be cast into a sealed mould.

### 6.7.1 Disc electrophoresis

Polyacrylamide was first introduced concurrently with a new electrophoretic method called "disc electrophoresis". This was first conducted in glass tubes, in which the separated protein bands constituted a series of discs. The method also embodied *disc*ontinuities in the buffer and gels used and so it may be described as discontinuous electrophoresis, abbreviated to disc. electrophoresis. Today, because of the better cooling and the fact that comparison of different samples is facilitated, disc. electrophoresis is generally conducted in vertical gel slabs. With this lay-out, the sample is applied to one end of the gel slab and so, unlike with a horizontal slab gel lay-out, only anions or cations, but not both, can be analysed at one time. The original method of Ornstein[6], is an anionic system (i.e. anions are analysed).

> The object of using discontinuous buffers and gels is to improve resolution by forming very narrow starting bands

In an anionic system, there is usually a common buffer cation, e.g. Tris$^+$, throughout. In the buffer compartments, a buffer with an anionic component having a pH-dependent electrophoretic mobility is used, e.g. glycine$^-$. Above its pI, the anodic mobility of glycine increases with pH as the proportion of glycine$^-$ ions increases.

$$\begin{array}{c} COOH \\ | \\ CH_2\text{-}NH_3^+ \end{array} \rightleftharpoons \begin{array}{c} COO^- \\ | \\ CH_2\text{-}NH_3^+ \end{array} \rightleftharpoons \begin{array}{c} COO^- \\ | \\ CH_2\text{-}NH_2 \end{array}$$

$$\xrightarrow{\hspace{6cm}}$$

Increasing pH

In the gels, an anionic component having a high, pH-independent mobility, e.g. $Cl^-$, is used.

Two gels are used; a large pore stacking gel which controls buoyancy-driven fluid flow (see p8) and a smaller pore running gel which imposes gel sieving. At the outset, both gels contain Tris-HCl buffer, but at different pH values. The buffer in the stacking gel is of lower pH than that in the running gel. A schematic view of the apparatus at the start of an electrophoresis run is shown in *Figure 85*. The samples are mixed with sucrose or glycerol to increase their density and are layered directly under the electrode buffer.

Upon application of the electrical field, a sharp interface develops between the high mobility so-called "leading ion", i.e. $Cl^-$, and the less mobile so-called "trailing ion", i.e. glycine, according to Kohlrausch's regulating functions[7]. In order to visualise the interface between the leading and trailing ions a small amount of a dye, such as bromophenol blue, may be added to the samples or to the upper electrode solution.

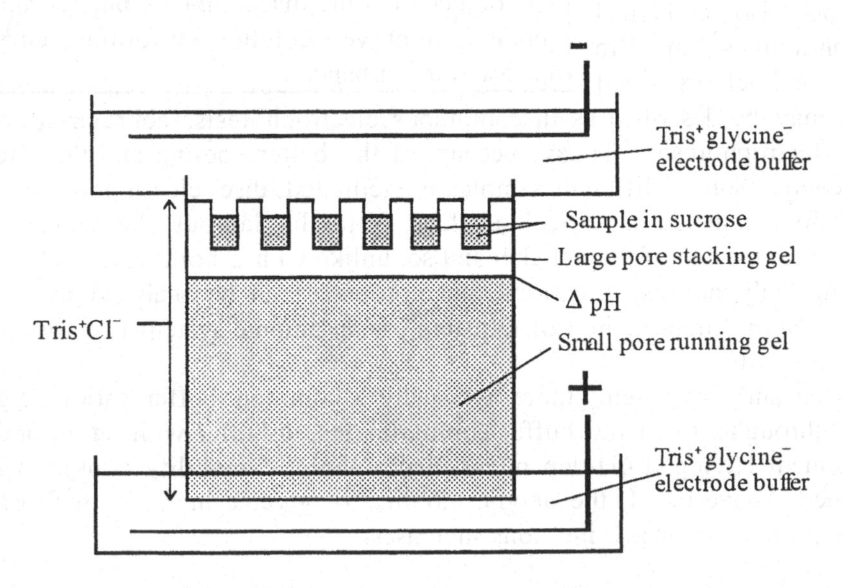


*Figure 85.* Experimental set-up for discontinuous PAGE.

As the interface moves downward, the protein molecules with mobilities intermediate between $Cl^-$ and glycine, will be swept up and concentrated into a very thin band. Within this band, individual proteins will become stacked in order of their mobilities, with those of highest mobility immediately next to the $Cl^-$ ions. During the later parts of this so-called stacking phase, therefore, all of the proteins will move downwards at the same speed, i.e. this could be called an isotachophoresis stage (iso = "the same", "tach" = speed).

The concentrated band of protein has a higher density than the surrounding buffer and the system would be subject to buoyancy-driven fluid flow (i.e. it would sink) were the system not stabilised by the stacking gel.

| The purpose of the large-pore stacking gel is to prevent buoyancy-driven fluid flow disturbances. |
|---|

The voltage gradient ($-dV/dx$) within the separating part of the apparatus will not be uniform but will have a stepwise decrement at the interface between the leading and trailing ions. If proteins are additionally present, the step will be comprised of a number of smaller sub-steps, corresponding to interfaces between the different proteins.

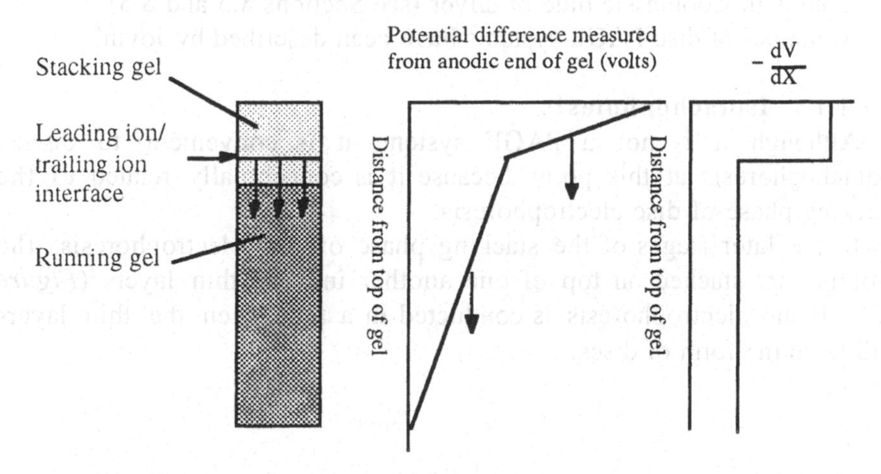


*Figure 86.* The voltage profile during the stacking phase of disc gel electrophoresis.

Arrows in *Figure 86* indicate the direction of movement of the interface and its associated voltage discontinuity. The voltage gradient is steeper behind the interface than in front. This follows from the fact that all the ions are moving at the same speed. For ions of a low mobility to move as fast as ions of a high mobility, they must be in a steeper voltage gradient. Any protein which falls behind the interface, say by diffusion, will find itself in an area with a steep voltage gradient and will thus be accelerated towards the interface. Conversely any protein which diffuses ahead of the interface will enter an area with a shallow gradient and will slow down and be overtaken by the moving interface. In this way the proteins are focused into thin layers in the interface.

When the interface reaches the junction between the stacking and running gels, two things happen. Firstly the pH increases, resulting in an increase in the mobility of the glycine trailing ion as a larger proportion will exist in the glycine$^-$ form at the higher pH. Secondly, the proteins encounter the sieving effect of the small pore running gel, which greatly increases the frictional resistance to movement. Together these result in the leading ion/trailing ion interface overtaking the stack of proteins, which are left to separate in a continuous linear voltage gradient according to their respective mobilities in the running gel.

The bromophenol blue dye remains with the interface, making its progress easily visible. When the interface reaches near the end of the gel the electric field can be switched off, in the sure knowledge that no proteins will have migrated further than the interface and that all will still be in the gel. The separated proteins can be visualised by staining, for example with Coomassie blue or silver (see Sections 8.5 and 8.6).

A number of disc PAGE systems have been described by Jovin[8].

#### 6.7.1.1 Isotachophoresis

Although it is not a PAGE system, it is convenient to discuss isotachophoresis at this point because it is conceptually related to the stacking phase of disc electrophoresis.

In the later stages of the stacking phase of disc electrophoresis, the proteins are stacked on top of one another in very thin layers (*Figure 87*). If the electrophoresis is conducted in a tube, then the thin layers will be in the form of discs.

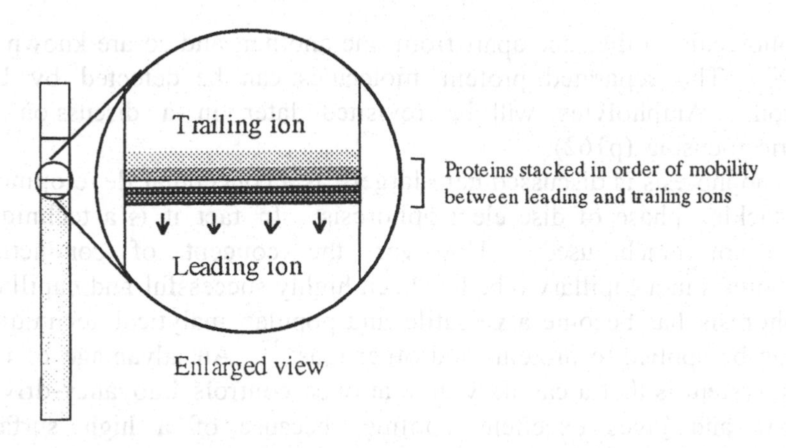


*Figure 87.* Proteins stacked as thin layers in the stacking phase of disc gel electrophoresis.

The different proteins are in fact separated from one another, although adjacent bands are touching, but this separation is not useful as the bands are so thin that it is impossible to distinguish between them. However, some improvement in the situation could be achieved by making the tube much thinner, so that a given amount of protein would occupy a greater length.

Ultimately, if the separation was carried out in a capillary tube, the bands would occupy a sufficient length that it becomes possible to distinguish the different bands. At each interface there is a step change in the voltage gradient, which corresponds to a change in resistance due to the fact that the proteins have different mobilities. This change in resistance at each interface corresponds to a change in heat production and this can be detected with a thermocouple detector. The output is a sharp peak as each interface passes the detector.

In this way the number of proteins present can be determined, but they cannot be identified. If one protein was missing, the others would simply close up and one fewer interface would be detected but it would be difficult to determine which protein was missing. This problem can be overcome by adding a mixture of "ampholytes" to the protein sample. Ampholytes are synthetic polyamino-polycarboxylic acids, in which the amino and carboxyl groups are randomly added to a carbon chain backbone - they are sold under a number of trade names, e.g. Ampholine, Servalyte, Bio-lyte. The result is a mixture of molecules with a range of pI values, resulting in a range of electrophoretic mobilities.

Because of their range of mobilities, there will always be some ampholyte molecules with mobilities intermediate between those of the different proteins and in isotachophoresis these will serve to space the protein molecules a distance apart from one another, and so are known as "spacers". The separated protein molecules can be detected by UV absorption. Ampholytes will be revisited later in a discussion of isoelectric focusing (p162).

Isotachophoresis is discussed here largely as a conceptual development of the stacking phase of disc electrophoresis. In fact it is a technique which is not much used. However, the concept of conducting electrophoresis in a capillary tube has been highly successful and capillary electrophoresis has become a versatile and popular analytical technique, which can be applied to proteins and other ions[9-17]. An advantage of the capillary system is that a capillary tube at once controls buoyancy-driven fluid flow and gives excellent cooling, because of a high surface area/volume ratio, so that high voltages can be used, giving rapid separations. A down-side to capillary electrophoresis is that the apparatus tends to be expensive.

## 6.8 SDS-PAGE

Sodium dodecyl sulfate-polyacrylamide gel electrophoresis (SDS-PAGE) was introduced in 1967 by Shapiro *et al.*[18] and has since become one of the most popular PAGE methods. The method is dependent in the first instance upon the interaction of the protein molecule with SDS. This is a detergent having a 12 carbon hydrophobic tail and a hydrophilic, sulfonic acid head group (**I**).

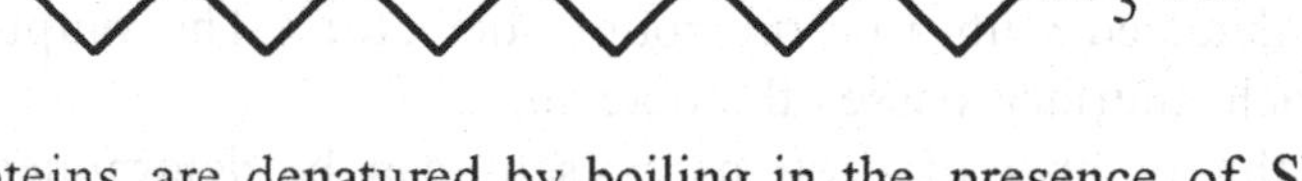

$SO_3^- \ Na^+$ **I**

Proteins are denatured by boiling in the presence of SDS. The SDS molecules interact with the proteins to give rod-like complexes, containing a constant ratio of *ca.* 1.4 mg of SDS per mg of protein[19]. At this level the negative charge on the SDS is sufficient to mask the charge on the protein and all proteins consequently have essentially the same charge/mass ratio and an anodic migration[20].

The lack of charge differences between different protein/SDS complexes means that the stacking phase will not be as effective as in conventional disc electrophoresis. However, Laemmli[21] has devised a convenient method whereby disc PAGE and SDS-PAGE can be conducted with the same set of reagents, simply with or without SDS. The Laemmli method has become one of the most popular PAGE methods in use today. However, the Laemmli method does not separate small proteins very well and an alternative SDS-PAGE method, using Tricine buffer, has been described by Shägger and Von Jagow[22]. The Tricine method gives uniquely good separation of proteins under 20 kDa (see Section 8.4).

As the intrinsic charge differences between proteins is masked by the SDS, separation of proteins is due solely to differences in size and hence the method can be used to determine molecular sizes[20]. A linear relationship between mobility and log *MW* obtains over a molecular weight range dependent upon the gel pore size. The gel can thus be standardised with proteins of known molecular weight and subsequently used to estimate the molecular weights of unknowns. Mobility is conveniently expressed as relative mobility $R_m$, i.e. mobility relative to the bromophenol blue tracker dye.

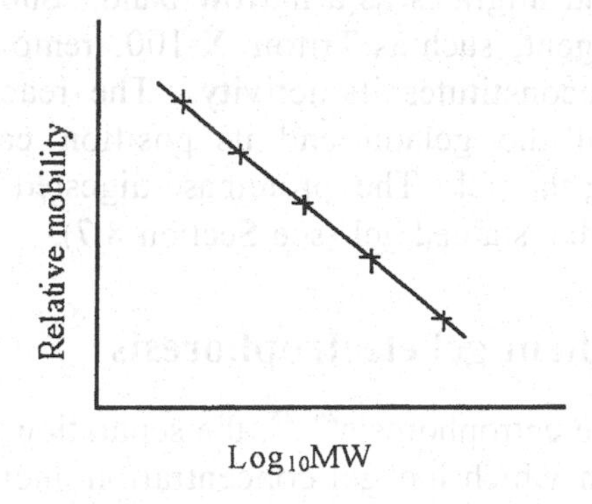


*Figure 88.* Standard curve for estimation of protein *MW*s by SDS-PAGE.

Some caveats apply to the estimation of molecular weights. If proteins are not reduced, then disulfide bridges may constrain the structure and prevent formation of the rod-like complexes. This will result in an incorrect apparent *MW*. On the other hand, this provides a way of detecting the existence of disulfide bridges. The glyco moiety of glycoproteins will also not bind SDS, and yet will contribute to the steric resistance of the molecule. In consequence, the molecular weights of glycoproteins tend to be overestimated. Finally, boiling in SDS not only denatures the protein but will dissociate the subunits of oligomeric proteins. The *MW*s obtained by SDS-PAGE will therefore be of the subunits and not of the intact protein.

### 6.8.1 An SDS-PAGE zymogram for proteinases

Zymography is a method for the detection of a specific enzyme among the bands separated by electrophoresis. The method usually relies on an enzyme specific reaction to generate a colour to reveal the position of the enzyme. Usually zymogram methods cannot be applied to SDS-PAGE as proteins are normally denatured in this technique. However, Heussen and Dowdle[23] have devised an ingenious SDS-PAGE zymogram method for proteinases. The method depends upon combining the proteinase-containing mixture with SDS, but without boiling the solution. In this form the SDS can apparently combine only partially with the protein, perhaps "tweaking" its conformation slightly and suppressing the activity of proteinases.

The SDS/protein mixture is separated by electrophoresis in a polyacrylamide gel containing a low concentration of gelatin (less than 1%). The proteinase/SDS complex does not bind to the gelatin, as would a free proteinase, and migrates as a narrow band. Subsequent incubation in a non-ionic detergent, such as Triton X-100, removes the SDS from the proteinase and reconstitutes its activity. The reactivated proteinase digests away part of the gelatin and its position can be detected by subsequently staining the gel. The proteinase-digested gelatin appears as a clear band on the blue stained gel (see Section 8.7).

## 6.9 Pore gradient gel electrophoresis

In pore gradient electrophoresis[24, 25], the separation is carried out in a direction in the gel in which the gel concentration increases and its pore size decreases. Proteins migrating along the gel concentration gradient will encounter an increasing frictional force, due to gel sieving, which will increasingly impede their progress. Eventually the proteins will cease migrating at a position where the gel pore size becomes smaller than their diameter. As different proteins will have different diameters, they will each reach a different position on the gel. By reference to standards, *MW*s can be calculated.

## 6.10 Isoelectric focusing

Of the separation methods based upon gross physical properties such as charge or size etc., isoelectric focusing[26,27] is one of the most discriminating. Only methods based upon some biological property, which requires a subtle stereospecific relationship, are more discriminating. Sometimes IEF can almost be **too** discriminating because multiple bands can be obtained from what is essentially the same glycoprotein species, with only small differences in glycosylation - a phenomenon known as "micro-heterogeneity". A downside of IEF is that it is a relatively expensive, because of the cost of the ampholytes required.

Proteins are ampholytes having a pH-dependent nett charge, which is positive at pH values below the protein's pI and negative at pH values above the pI.

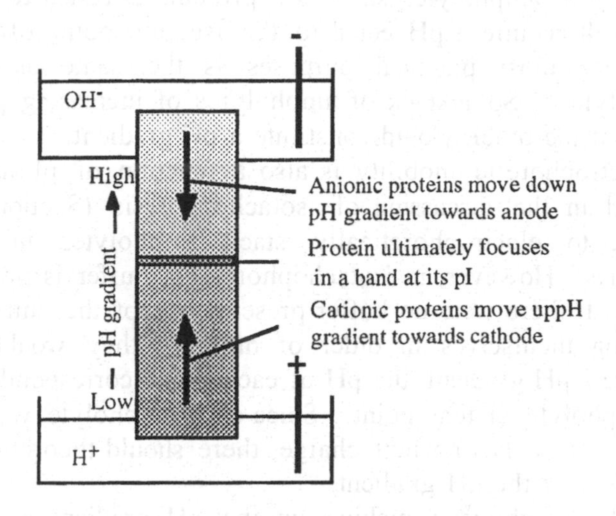


*Figure 89.* Schematic view of isoelectric focusing.

If proteins are distributed throughout a solution in a pH gradient, then upon application of an electrical potential across the gradient, with the anode at the low pH end, molecules in the low pH zone (which will be positively charged) will migrate to the cathode. Their migration will take them through zones of increasing pH, as a consequence of which they will gradually lose their positive charge and their rate of migration will slow down. Conversely, molecules in the high pH zone will be negatively charged and will consequently migrate, through zones of decreasing pH, towards the anode. When each protein reaches a position where the pH is equal to its pI, it will lose all of its charge and its migration will cease. After a sufficient time, therefore, the respective molecules will in consequence become "focused" at their isoelectric points. In this way, a mixture of proteins can be separated, as each will focus at a characteristic pI. Furthermore, the pI values can readily be measured in this way. In practice, three difficulties must be overcome:

- A stable, uniform pH gradient must be established and maintained.
- The system must be stabilised against disturbances due to buoyancy-driven fluid flows.

- A system must be devised for the measurement of the pH gradient and for determination of the positions of the focused bands.

### 6.10.1 Establishing a pH gradient

If a pure ampholyte, such as a protein, is added to pure water, the water will acquire a pH equal to the isoionic point of the ampholyte, which for most practical purposes is the same as the pI of the ampholyte[28]. So, a stack of ampholytes of increasing pI, arranged one on top of the other, would constitute a pH gradient.

Electrophoretic mobility is also a function of pI and, as has been outlined in the discussion of isotachophoresis (Section 6.7.1.1), it is possible to electrophoretically stack ampholytes in order of their mobilities. However, in isotachophoresis a buffer is present to control the pH. If there were no buffer present, except the ampholytes, then in arranging themselves in order of mobility they would simultaneously generate a pH gradient, the pH at each point corresponding to the pI of the ampholyte at that point. Since each ampholyte would finally be at its pI, where it has no nett charge, there should theoretically be no nett movement of the pH gradient.

If the ampholytes making up the pH gradient were proteins, the gradient would have a few steps (as many as there are proteins), but these steps would tend to be quite large (*Figure 90*). However, if synthetic, randomly substituted, polyamino-polycarboxylic acid ampholytes were used (see Isotachophoresis, p154), then there would be a very large number of very small steps, which in effect gives a smooth pH gradient. A protein introduced into such a gradient will cause a plateau to be formed at its pI, the length being proportional to the amount of protein (*Figure 91*).

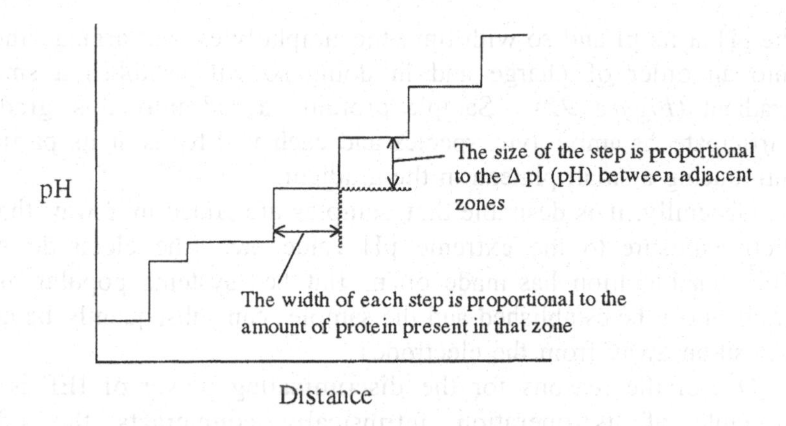
pH
The size of the step is proportional to the Δ pI (pH) between adjacent zones
The width of each step is proportional to the amount of protein present in that zone
Distance

*Figure 90.* A pH gradient constructed from a stack of seven proteins.

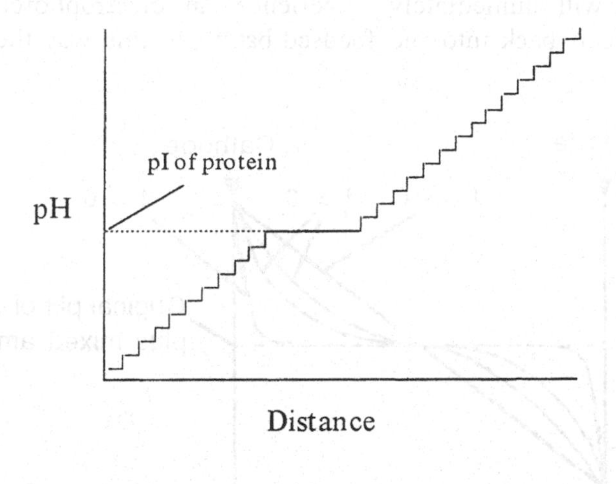
pI of protein
pH
Distance

*Figure 91.* A pH gradient constructed from randomly synthesised polyamino-polycarboxylic acid ampholytes (and containing one protein).

The anode solution is acidic and the cathode solution is basic. Consequently, ampholyte molecules immediately in contact with the anode solution will be positively charged and those in contact with the cathode solution will be negatively charged. At zero time the pH distribution across the apparatus will be low at the anode end and high at the cathode end, but with a central plateau, corresponding to the pH of the sample plus mixed ampholytes (*Figure 92*). When the electrical potential is applied across the electrodes, ampholytes will move to the anode or cathode, depending upon their charge, until they reach a pH at which they will have zero charge. Each ampholyte species will establish the pH at its pI and so with time the ampholytes will arrange themselves into an order of charge and in doing so will establish a smooth pH gradient (*Figure 92*). Sample proteins added into this gradient will participate as ampholyte species and each will focus at its particular pI, introducing a small plateau in the gradient.

Generally, it is desirable that samples are added in a way that avoids their exposure to the extreme pH values near the electrode solutions. This consideration has made open, flat bed systems popular as the pH gradient can be established and the samples can subsequently be applied at a position away from the electrodes.

One of the reasons for the discriminating power of IEF is that the principle of its operation intrinsically counteracts the effects of diffusion, which in other methods is responsible for the broadening of bands with time. Any protein which diffuses out of a focused band will enter a region of different pH where it will acquire a charge. In consequence it will immediately experience an electrophoretic force tending to move it back into the focused band. In this way the band is kept focused.

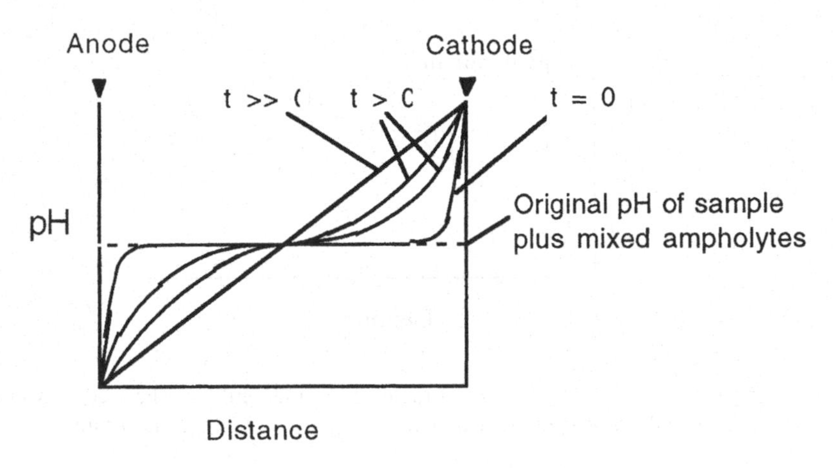


*Figure 92.* Time course of the establishment of a pH gradient in IEF.

The steepness of the pH gradient, and its consequent resolving power, can be altered by using ampholytes covering different pH ranges. The range pH 3-10 is used first to get an overview of where the proteins focus and an appropriate choice of ampholytes can be made for a second round of focusing over a smaller range, say pH 5 to 7. The smaller the pH range covered, the greater will be the resolution.

### 6.10.2 Control of buoyancy-driven fluid flow

Buoyancy-driven currents are induced in fluids by the effects of gravity upon fluids in which a more dense portion of the fluid occurs above a less dense portion. In IEF they are caused by i) heat effects, which reduces the density of the heated solution, hereby inducing convection and, ii) from the fact that focused protein bands are more dense than the surrounding ampholyte solutions and will thus tend to sink. Practical systems consequently require some way of obviating these buoyancy-driven fluid flow disturbances.

Since buoyancy-driven fluid flow depends upon differences in density, the effect of gravity and a consequent movement of fluid, to avoid it any one or more of these three elements can be targeted. Thus, an early approach was to conduct IEF in a sucrose gradient, thereby pre-imposing a density gradient which would damp out any buoyancy-driven disturbances. Alternately, because buoyancy-driven disturbances are a consequence of fluid movement, another approach is to have a system where the liquid cannot move, such as in a gel, and a popular modern method is to conduct IEF in a gel slab. To eliminate the effects of gravity, IEF may be conducted in a rotating tube, so that the gravity vectors cancel out with time, or, at somewhat greater expense, the experiment may be conducted under microgravity conditions in a free-falling spaceship in orbit around the Earth. In these cases a gel is not required.

### 6.10.3 Applying the sample and measuring the pH gradient

In any practical system, all of the attendant problems must be solved simultaneously. Thus, in addition to controlling buoyancy-driven fluid flow, the system must allow for sample application - avoiding the pH extremes - and for measurement of the pH gradient and the position of focused bands, after the separation. Different systems are suitable for analytical and preparative purposes and one of each will be briefly described, by way of illustration.

#### 6.10.3.1 An analytical IEF system

The most common analytical system in use at present is the thin slab gel system. In this a thin layer of gel - either large pore polyacrylamide or agarose - containing an appropriate ampholyte mixture, is cast onto a backing sheet of Gel-Bond®, a product of FMC Inc. The sheet is positioned on top of a template on a cooling block. The template marks sample tracks, and serves as a guide to the application of the sample.

Wicks, made of several layers of filter paper, are impregnated with the appropriate acid or base electrode solution and placed on top of the gel at each end (*Figure 93*). When the apparatus is sealed, electrodes contact the filter paper wicks, thereby closing the electrical circuit (this description is based on the Pharmacia Biotech Multiphor apparatus).

The pH gradient is allowed to become established before the sample is applied. Sample is applied to the gel by carefully laying small rectangles of filter paper, impregnated with sample solution, on the gel over a track mark on the template. If necessary, the same sample solution can be applied at different positions, i.e. at different pH values, on the gel.

Samples are drawn out of the filter paper and into the gel by electrophoresis and by diffusion. When this has happened the potential is switched off and the sample applicator papers are removed, so that they do not subsequently distort the electrical field.

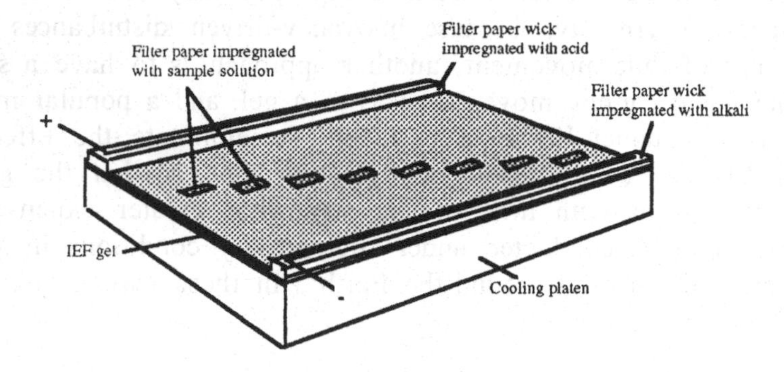


*Figure 93.* Sketch of apparatus used for analytical flat bed IEF.

After focusing is complete, the gel is removed and stained to reveal the position of the protein bands. Ampholytes are polyamino compounds and at most pH values they will react with and precipitate dyes such as Coomassie blue. To obviate this, the ampholytes can be removed by washing the gel in trichloracetic acid before staining. A simpler method, however, is to use the principle of the Bradford assay for protein (see Section 3.3.5), i.e. to stain with Coomassie blue G-250 in acid solution[29]. Using this approach only the proteins are stained and the ampholytes do not interfere.

In order to determine the pI values of the separated proteins, it is necessary to measure the pH gradient. This can be done, after focusing but before staining, by using a surface-probing pH electrode. However, as with other techniques, once the values for a few proteins have been established, these can be used as standards, and the values of unknowns can be determined by interpolation. Standard proteins, of known pI values can thus be run in parallel with unknowns - on the same gel but in different tracks - and a standard curve of pH *vs* distance can be constructed from the standards and used to determine the pI values of the unknowns. A table of pI values of standard proteins has been published by Chambers and Rickwood[30], and pI calibration kits are commercially available.

#### 6.10.3.2 Preparative IEF

Preparative IEF[31] differs from analytical IEF in that it is done on a much larger scale, i.e. with much more sample and, as the products sought are active protein fractions, provision must be made for recovery of the separated fractions after IEF. The central problem with preparative IEF is stabilisation of the system against buoyancy-driven fluid flows during the focusing process while still being able to elute the separated components at the end of the process. Different approaches to the solution of this problem have been adopted by different authors and by the makers of commercially available apparatus. The most common approach is to conduct the focusing in a buoyancy-flow-stabilised liquid phase. With the proteins and ampholytes being recovered in solution, measurement of the protein concentration and pH of each fraction is relatively straightforward. An alternative approach, suited to smaller-scale preparative uses, is to conduct the focusing in a slab gel, to cut out the focused band and electro-elute this[32,33]. In this case the pH gradient would be measured as described for analytical IEF.

The first approach tried, and commercialised by Pharmacia, was to conduct the focusing in a sucrose gradient, which could be eluted from the apparatus at the end of the experiment. An ingenious, but expensive, apparatus had provisions for cooling the annular focusing column on both its inner and outer surfaces, and valves to isolate the electrode solutions during the elution phase.

An approach used by LKB (which has since merged with Pharmacia and been incorporated into Amersham Biosciences) was to use a flat bed of granulated gel. After focusing, the gel bed could be divided into a number of segments, the gel from each segment being scooped out and packed into a mini-column from which the focused protein and ampholytes could be eluted.

In the Bio-Rad Rotofor™ apparatus, buoyancy-driven fluid flow is controlled by conducting the focusing in a horizontal column which is rolled at 1 rpm about its central axis. Rolling serves to negate the effects of gravity and enables the proteins to be focused in free solution. A central ceramic cold finger removes heat. The column is divided into 20 segments by polyester membranes. After focusing, solution in the segments can be rapidly eluted from the side of the column into 20 test-tubes, without mixing. The Rotofor is currently a favoured apparatus for preparative IEF[30].

Although good results can be obtained by preparative IEF, and for many separations it may be the only practicable method, the technique is constrained by the high initial cost of the apparatus and the cost of the ampholytes consumed in each experiment.

## 6.11 2-D Electrophoresis

In 2-D electrophoresis, proteins are separated in the first dimension, according to their isoelectric points, by IEF, and in the second dimension, according to their molecular weights, by SDS-PAGE[34,35]. The first, IEF, stage is conducted in a long, thin, spaghetti-like gel. This gel must be of large pore-size so that focusing is possible, but this makes it very soft and fragile. For the second stage the long, thin, gel is transferred to the origin of a slab gel, sealed in position and an SDS-PAGE separation is carried out. The result, after staining, is a number of spots distributed in two dimensions over the slab gel. Because the first-stage gel is of a small diameter, only a small amount of ampholytes is consumed per run and so 2-D electrophoresis is relatively inexpensive.

Although conceptually simple, it took a number of years for the method to be developed to its present stage of practicability. One of the technical difficulties to be overcome was the handling and transferral of the first-stage gel - soft spaghetti is not the easiest type of material to handle!

## 6.12 Non-linear electrophoresis

Electrophoresis can be visualised by an analogy. Imagine spherical glass marbles rolling down a slope in a tank of syrup. The profile of the slope could be described by plotting contour lines. The marbles would move very slowly because of the viscosity of the syrup and so they would essentially acquire no momentum. As a result they would always move in the direction of the slope, even if this was to change its direction, i.e. the locus of any one marble would always be at right angles to the contour lines. The locus would reflect the direction of the nett force vector acting upon each marble and the magnitude of the force would be inversely proportional to the spacing between the contour lines.

In electrophoresis, the equivalent of contour lines would be *isovoltage contours* and the loci of ions undergoing electrophoresis would correspond to *field lines*, which are always at right angles to the isovoltage contour lines.

In most forms of electrophoresis, the field lines are straight and the isovoltage contours are evenly spaced. In the marbles-in-syrup analogy, this is equivalent to the marbles moving down a flat plane surface, inclined at an angle to the horizontal. However, deviating from this "straight and narrow" approach can be instructive and one can pose a number of "what if?" questions to probe one's own insight into electrophoresis, e.g.

- What if the gel was of increasing cross-section, i.e. either conical or wedge-shaped?
- What if the gel was not straight, but went through a 90° bend?
- What if the gel had a hole cut in it (say if there was a pillar passing through the gel)?

These questions have been explored theoretically and empirically by Dennison *et al.*[36,37].

If the gel were conical or wedge-shaped, the voltage gradient ($dV/dx$) would be steeper at the narrow end and shallower at the wider end. The nett result is that the migration of slower ions would be increased, as they would spend more time in the steep part of the gradient, whereas that of faster ions would be slowed as they would reach the shallower parts of the gradient more quickly. Normally, in gel electrophoresis the mobility of ions is a logarithmic function of their molecular weight, so that small ions are separated better than larger ions. In a wedge-shaped gel the relationship is made more linear[35]. Although this result has not found much utility in the separation of proteins, wedge gels have proved useful in the separation of nucleic acids. In DNA sequencing gels, a greater number of bases can be sequenced in a wedge gel compared to a straight-sided slab gel[38-40], and the separation of plasmids is also improved in wedge-shaped gels[41]. The wedge-gel concept has been extended to preparative electrophoresis by Rolchigo and Graves[42], and to IEF by Pflug[43].

An empirical analysis of electrophoresis around corners was done by Dennison *et al.*[37] (*Figure 94*). Around a corner, the isovoltage contours remain as straight lines but become closer together on the inside of the corner than on the outside, rather like the steps in a spiral staircase. As a result, the voltage gradient is steeper on the inside of the corner and the proteins on this side will accelerate when negotiating the corner. By contrast, the isovoltage contours are further apart on the outside of the corner and the proteins will slow down. The nett result is that the protein band will become skewed as it rounds the corner (*Figure 95*).

Negotiating a second corner in the opposite direction does not "undo" the distortion[37].

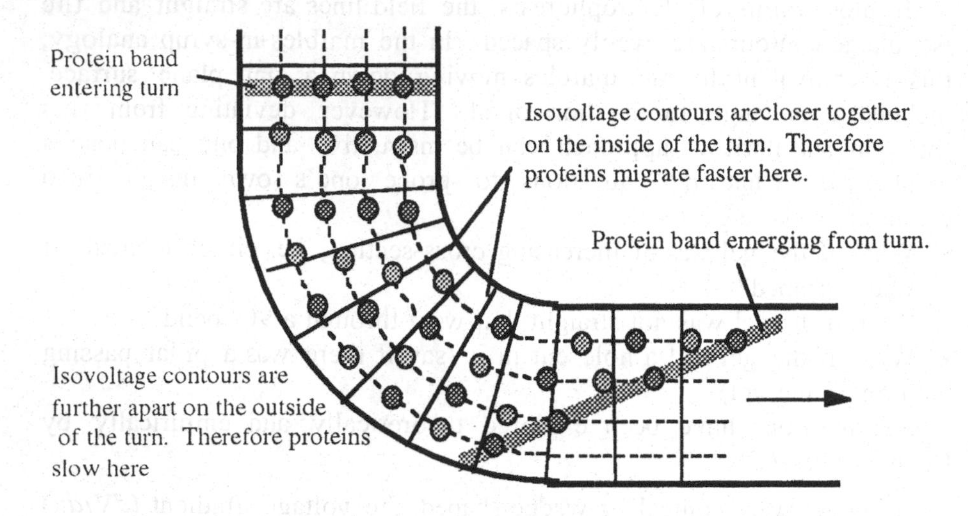
Protein band
entering turn
Isovoltage contours arecloser together
on the inside of the turn. Therefore
proteins migrate faster here.
Protein band emerging from turn.
Isovoltage contours are
further apart on the outside
of the turn. Therefore proteins
slow here

*Figure 94.* The effect of a bend in the gel upon electrophoretic behaviour in PAGE. From Dennison *et al.*[27].

Electrophoresis around a circular obstruction in the gel is shown in *Figure 95*. The isovoltage contour lines (shown as solid lines) can easily be visualised, remembering the fact that they are always at right angles to the field lines (shown as dotted lines). An interesting conclusion was drawn from these experiments, i.e. that the equations describing the behaviour of ions undergoing electrophoresis are identical to those describing the irrotational flow of an ideal incompressible fluid. Ideal fluid behaviour was previously considered to be only an abstract concept; in an ideal fluid the molecules have a viscosity of zero.

If one has a playful nature, the shapes of the voltage gradient surfaces can be visualised on a visit to the beach. Pouring loose beach-sand down a slope, in which a circular obstruction has been placed, will form a contoured surface around the obstruction. In one's mind's eye one could imagine this surface in the tank of syrup and visualise the movement of marbles down the surface - it will be found to be identical to the behaviour of ions around the circular obstruction shown in *Figure 95*!

To summarise, the following points can be noted about non-linear electrophoresis:-

- Field lines pass smoothly around corners or obstructions and resemble "streamlines".
- Ions will migrate along field lines, i.e. the field lines will represent the loci of migrating ions.

- Isovoltage contour lines are always at right angles to the field lines.
- Ions will migrate quickly where the isovoltage contour lines are close together (i.e. where the voltage gradient is steep) and slowly where the isovoltage contour lines are far apart (i.e. where the voltage gradient is shallow).

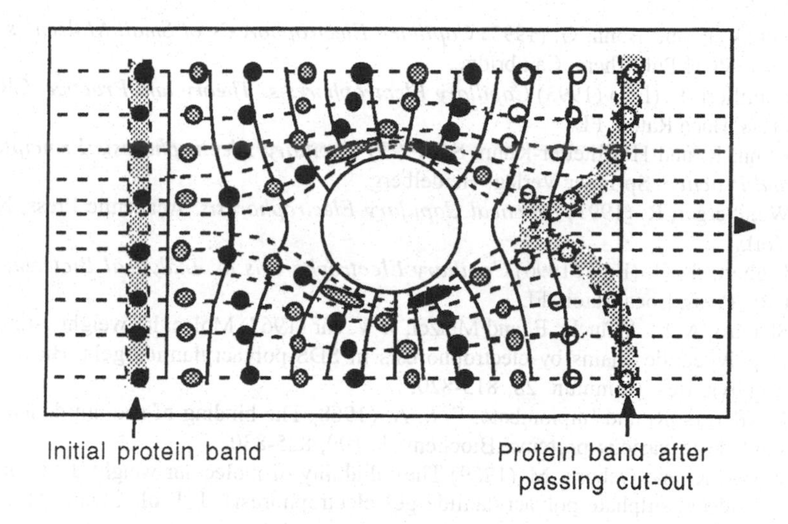


*Figure 95.* The effect of a circular hole in an electrophoresis gel.

## *References*

1. Garnett, P. and Garnett, P. (1996) Workshop 1.8. 14th International Conference on Chemical Education, Brisbane, Australia.
2. Kyte, J. (1995) in *Structure in Protein Chemistry*. Garland Publishing Inc. New York, pp27-33.
3. Hjertén, S. (1961) Agarose as an anticonvection agent in zone electrophoresis. Biochim. Biophys. Acta 53, 514-517.
4. Smithies, O. (1955) Zone electrophoresis in starch gels: group variations in the serum proteins of normal human adults. Biochem J. 96, 595-606.
5. Ferguson, K. A. (1964) Starch gel electrophoresis - application to the classification of pituitary proteins and polypeptides. Metabolism 13, 985-1002.
6. Ornstein, L. (1964) Disc electrophoresis. I. Background and theory. Ann. N. Y. Acad. Sci. 121, 321-349.
7. Jovin, T. M. (1973) Multiphasic zone electrophoresis. I. Steady-state moving-boundary systems formed by different electrolyte combinations. Biochemistry 12, 871-879.
8. Jovin, T. M. (1973) Multiphasic zone electrophoresis. II. Design of integrated discontinuous buffer systems for analytical and preparative fractionation. Biochemistry 12, 879-890.

9. Grossman, P. D. and Colburn, J. C. (Eds.) (1992) *Capillary Electrophoresis: Theory and Practice*. Academic Press, New York.
10. Li, S. F. Y. (1992) *Capillary Electrophoresis:Principles, Practice and Applications*. Elsevier, Amsterdam.
11. Wiktorowicz, J. E. (Ed.) (1992) *Capillary Electrophoresis*. Academic Press, New York.
12. Guzman, N. A. (Ed.) (1993) *Capillary Electrophoresis Technology*. Dekker, New York.
13. Jandik, P. and Bonn, G. (1993) *Capillary Electrophoresis of Small Molecules and Ions*. VCH Publishers, Cambridge.
14. Camilleri, P. (Ed.) (1993) *Capillary Electrophoresis: Theory and Practice*. CRC Press, Boca Raton, FL.
15. Kuhn, R. and Hoffdtetter-Kuhn, S. (1993) *Capillary Electrophoresis:Principles and Practice*. Springer Verlag, Heidelberg.
16. Weinberger, R. (1993) *Practical Capillary Electrophoresis*. Academic Press, New York.
17. Righetti, P. G. (Ed.) (1996) *Capillary Electrophoresis in Analytical Biotechnology*. CRC Press, Boca Raton, FL.
18. Shapiro, A. L., Viñuela, E. and Maizel, J. V. Jnr (1967) Molecular weight estimation of polypeptide chains by electrophoresis in SDS-polyacrylamide gels. Biochem. Biophys. Res. Commun. 28, 815-820.
19. Pitt-Rivers, R. and Impiombato, F. S. A. (1968) The binding of sodium dodecyl sulphate to various proteins. Biochem. J. 109, 825-830.
20. Weber, K. and Osborn, M. (1969) The reliability of molecular weight determinations by dodecyl-sulphate-polyacrylamide gel electrophoresis. J. Biol. Chem. 244, 4406-4412.
21. Laemmli, U. K. (1970) Cleavage of structural proteins during the assembly of the head of bacteriophage T. Nature 277, 680-685.
22. Shägger, H. and von Jagow, G. (1987) Tricine-sodium dodecyl sulfate-polyacrylamide gel electrophoresis for the separation of proteins in the range from 1-100 kDa. Anal. Biochem. 166, 368-379.
23. Heussen, C. and Dowdle, E. B. (1996) Electrophoretic analysis of plasminogen activators in polyacrylamide gels containing sodium dodecyl sulfate and copolymerized substrates. Anal. Biochem. 102, 196-202.
24. Margolis, J. and Kenrick, K. G. (1967) Polyacrylamide gel electrophoresis across a molecular sieve gradient. Nature (London) 214, 1334-1336.
25. Slate, G. G. (1968) Pore-limit electrophoresis on a gradient of polyacrylamide gel. Anal. Biochem. 24, 215-217.
26. Vesterberg, O. (1971) Isoelectric focusing of proteins. Methods Enzymol. 22, 389-412.
27. Wrigley, C. (1971) Gel electrofocusing. Methods Enzymol. 22, 559-564.
28. Kyte, J. (1995) in *Structure in Protein Chemistry*. Garland Publishing Inc. New York, pp25-27.
29. Blakesley, R. W. and Boezi, J. A. (1977) A new staining technique for proteins in polyacrylamide gels using Coomassie Brilliant Blue G250. Anal. Biochem. 82, 580-582.
30. Chambers, J. A. A. and Rickwood, D. (1993) in *Biochemistry Labfax*. βios Scientific Publishers, Oxford, pp75-76.
31. Garfin, D. E. (1990) Isoelectric focusing. Methods Enzymol. 182, 459-477.

32. Ziola, B. R. and Scraba, D. G. (1976) Recovery of SDS-proteins from polyacrylamide gels by electrophoresis into hydroylapatite. Anal. Biochem. 366-371.
33. Polson, A. von Wechmar, M. B. and Moodie, J. W. (1978) Preparative immunoabsorption electrophoresis. Immunol. Commun. 7, 91-102.
34. Garrels, J. I. (1983) Quantitative two-dimensional gel electrophoresis of proteins. Methods Enzymol. 100, 411-423.
35. Dunbar, B. S., Kimura, H. and Timmons, T. H. (1990) Protein analysis using high-resolution two-dimensional polyacrylamide gel electrophoresis. Methods Enzymol. 182, 441-459.
36. Dennison, C., Lindner, W. A. and Phillips, N. C. K. (1982) Nonuniform field gel electrophoresis. Anal. Biochem. 120, 12-18 (1982).
37. Dennison, C., Phillips, A. M. and Nevin, J. M. (1983) Nonlinear gel electrophoresis: an analogy with ideal fluid flow. Anal. Biochem. 135, 379-382.
38. Ansorge, W. and Labiet, S. (1984) Field gradients improve resolution on DNA sequencing gels. J. Biochem. Biophys. Methods 10, 237-243.
39. Bonicelli, E., Simeone, A., deFalco, A., Fidanza, V. and La Volpe, A. (1983) An agarose gel resolving a wide range of DNA fragment lengths. Anal. Biochem. 134, 40-43.
40. Olsson, A., Moks, T., Uhlen, M. and Gaal, A. B. (1984) Uniformly spaced banding pattern in DNA sequencing gels by use of field-strength gradient. J. Biochem. Biophys. Methods 10, 83-90.
41. Rochelle, P. A., Day, M. J. and Fry, J. C. (1986) The use of agarose wedge-gel electrophoresis for resolving both small and large naturally occurring plasmids. Lett. Appl. Micro. 2, 47-51.
42. Rolchigo, P. M. and Graves, D. J. (1988) Analytical and preparative electrophoresis in a non-uniform electric field. AIChE J. 34, 483-492.
43. Pflug, W. (1985) Wedge-shaped ultrathin polyacrylamide and agarose gels for isoelectric focusing. Electrophoresis 6, 19-22.

## 6.13 Chapter 6 study questions

1. Express in words the effect, in the electrophoresis of a protein, of, a) increasing the steepness of the voltage gradient, b) increasing the charge on the protein.
2. A protein is moving by electrophoresis through a buffer which does not contain sucrose. Explain what would happen if the protein migrated into a region of buffer containing, say, 20% sucrose.
3. If the cross-sectional area of a gel is decreased, what will be the effect on its electrical resistance?
4. In an electrophoresis system, at equal ionic strength, the conductivity of a smaller ion is greater/smaller than that of a smaller ion and this will cause the protein ions to have a lower/higher mobility (select the correct word in each case).

5. Towards which electrode does the buffer tend to flow, in most forms of electrophoresis, and what is this due to?
6. Answer True or False. i) Water can siphon through agarose gel, ii) SDS-PAGE is a useful preparative technique, iii) In disc-PAGE the anode must be in the lower vessel.
7. Name one advantage and two disadvantages of polyacrylamide over starch, as a medium for electrophoresis.
8. What will be the pH of a solution of a pure protein dissolved in pure water?
9. In disc-PAGE, what is the function of a) the stacking gel?, b) the running gel?
10. An appropriate standard curve applicable to SDS-PAGE, is ________ *vs* ________?
11. In relation to that measured by MEC, the *MW* measured by SDS-PAGE will always be smaller, smaller-or-equal, larger, larger-or-equal?
12. What are "ampholytes"?
13. What is the object of using discontinuous gel and buffer systems in disc-PAGE?
14. List the problems which must be simultaneously addressed in the design of a practical IEF system.
15. In a PAGE experiment in which the gel slab is 10 cm long, a potential of 300 V is applied across the gel, giving a current of 30 mA. a) What is the voltage gradient? b) If a further, identical, gel slab is run in parallel with the first and the current is kept constant at 30 mA, what will the new voltage gradient be? c) Qualitatively, what would happen to the voltage gradient if a smaller buffer ion was used?
16. A standard mixture, comprised of six proteins of known MW (12.9, 16.6, 21.6, 31.2, 40.5, 59.4 kDa), plus bromophenol blue, was subjected to SDS-PAGE, in parallel with a purified unknown protein. Under non-reducing conditions the standard mixture gave six stained bands, 1.1, 1.8, 2.6, 3.8, 4.4 and 5.0 cm from the origin, while the unknown yielded one band 1.5 cm from the origin. The bromophenol blue, in each gel, migrated 6.1 cm. Under reducing conditions, the standard mixture gave seven stained bands, 1.2, 2.7,

3.5, 4.0, 4.6, 4.8 and 5.2 cm from the origin, while the unknown yielded two bands, 3.1 and 3.7 cm from the origin, respectively. The bromophenol blue, in this case had migrated 6.4 cm in each gel. a) Interpret these results, in quantitative terms, b) Explain what you understand by, "reducing" and, "non-reducing" conditions.

17. Can water move through a gel under the influence of gravity?
18. Can water move through a gel under the influence of an electric field? Explain.
19. How can the pore size of a gel be varied?
20. What is meant by a, "macroreticular" gel?
21. You are running an SDS-PAGE experiment. In a previous experiment you observed that, at a potential of 90 volts, the bromophenol blue marker migrated 35 mm in 1 h. In your present experiment, in the same gel, you are using a potential of 150 volts and the bromophenol blue has 50 mm to go, before reaching the end of the gel. You wish to leave the lab, to go for supper: how long can you afford to be away?
22. Joe was conducting an electrophoresis experiment in a flat bed system, with the ends of the gel connected to the electrode buffers with filter paper wicks. Because he thought it wouldn't matter, Joe had made the filter paper wicks wider than the gel, as shown in the sketch below (only one end of the gel is shown). By an analysis of the field lines and the isovoltage contours, explain why this arrangement would lead to the protein bands becoming curved as shown in the sketch.

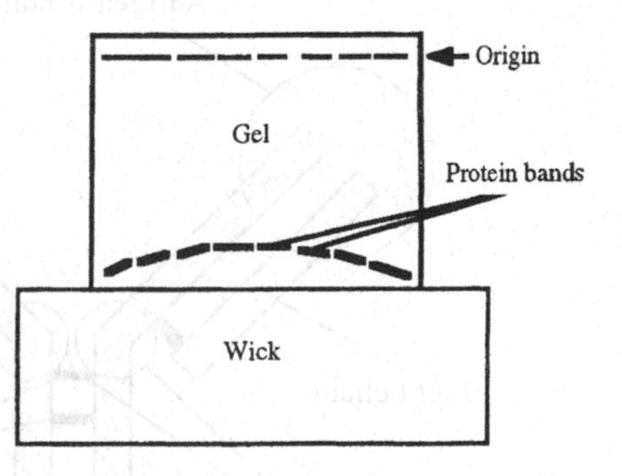

# Chapter 7

## Immunological methods

Antibodies are proteins made by the immune system of animals as part of a defence system against infection by foreign organisms. The immune system must be able to distinguish between "self" and "non-self", and to eliminate the latter, and antibodies play a role in this process. Because of their specificity and versatility, antibodies are also very useful reagents in the identification and analysis of proteins and it is largely in this light - as useful reagents - that antibodies will be considered in this chapter.

### 7.1 The structure of antibodies

Antibodies are a class of blood proteins known as γ-globulins. The most common γ-globulin is the IgG type, which consists of four peptide chains, two heavy and two light, held together by disulfide bonds. A schematic sketch of the structure is shown in *Figure 96*.

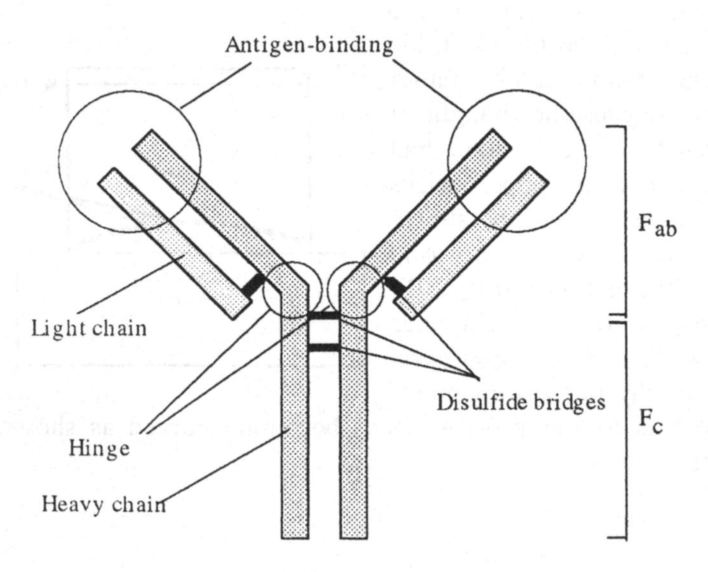


*Figure 96.* A simplified schematic representation of the structure of an IgG antibody.

The IgG molecule can be cleaved in the hinge region by papain, to yield three fragments, two $F_{ab}$ fragments and one $F_c$ fragment. "$F_{ab}$" stands for "fragment, antigen binding" and this reflects the fact that the outer aspect of the heavy and light chains, in the $F_{ab}$ fragment, contain so-called hypervariable regions which constitute the antigen binding site. The hypervariable regions are different in the IgG molecules synthesised by different plasma cell clones, but will be identical in every molecule originating from a single clone of plasma cells (See p174). The hypervariable regions are constituted of loops at one end of a β-barrel structure[1]. Such loops can vary without disturbing the underlying stability of the barrel structure. "$F_c$" stands for "fragment, crystallisable" and this reflects the fact that the $F_c$ fragment is invariant and therefore can be crystallised, even when it is derived from a polyclonal antiserum.

In mammals, antibodies are transferred to the neonate in the form of colostrum, which is the first milk produced in the early post-partum stage. In birds the transfer of antibodies occurs via the egg yolk and egg yolk thus provides a convenient source from which antibodies may be isolated. The antibodies from egg yolk have a structure similar, but not identical, to IgG and are known as IgY antibodies.

## 7.2 Antibody production

The immune system is designed to ward off "foreign" invaders. Injection of a foreign molecule, usually a protein, into an animal elicits an immune response, which includes the production of antibodies which react with the foreign protein and target it for removal from the system. The foreign protein in this context used to be called an **antigen**, from **anti**body **gen**erator, but the terminology has changed and now the molecule which elicits antibody production is called an **immunogen** and the molecule with which the antibody reacts is called an **antigen** (*Figure 97*).

| **Immunogen** - elicits antibody production |
| --- |
| **Antigen** - reacts with antibody |

The terminology changed when it was realised that although the immunogen and the antigen are often the same molecule, sometimes they are not. Also, some molecules are antigenic (i.e. they react with antibodies) but are not immunogenic (i.e. they will not elicit antibody production when injected into an animal). The current hypothesis takes account of the fact that the immunogen and the antigen may or may not be the same molecule.

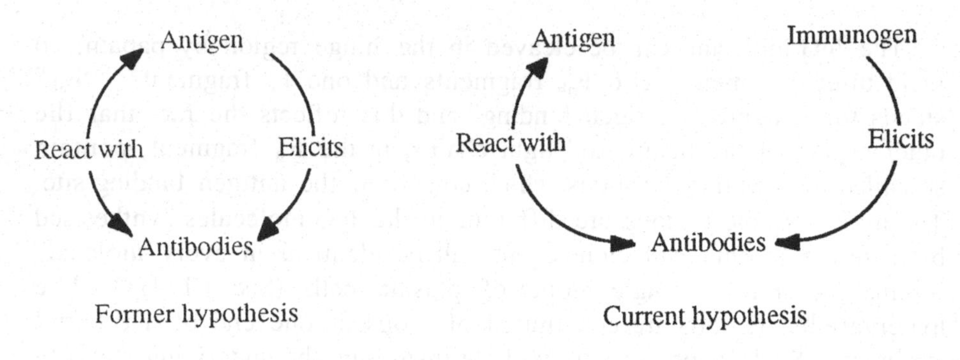


*Figure 97.* The relationship between immunogens, antibodies and antigen.

Many small molecules will not by themselves elicit antibody production but when conjugated to a larger molecule they will elicit antibodies, which will react with the unconjugated small molecule. Such small molecules are known as *haptens*. The cut-off in size is not absolute and experience with peptide antibodies (antibodies elicited by peptide immunogens) suggests that it may be that smaller molecules simply take *longer* to elicit antibodies. For example, free peptides are often able to elicit antibodies, but the antibody titre takes longer to rise than when a peptide conjugated to a carrier protein is used as the immunogen.

> *Haptens* are small molecules which can react with specific antibodies but which, by themselves, cannot elicit antibody production.

A single injection of an immunogen is not optimally effective in eliciting antibody production. In a natural infection, molecules from the infecting agent will leak from the site of infection to become exposed to the immune system in small amounts, but over an extended time. To elicit antibodies, this natural process must be mimicked. One way would be to inject a small amount of immunogen at a time and to repeat the injection many times over a time period. This would work, but it would be tedious and would subject the animal to unnecessary trauma, which is ethically unacceptable. Another way would be to make an emulsion of the aqueous antibody solution with an adjuvant oil and inject this subcutaneously or intramuscularly. The emulsion is made by a process known as *trituration*. The injected emulsion would exist at a focal site, mimicking a natural infection, and would break down slowly over time, thereby slowly releasing the immunogen and exposing it to the immune system over a

> An *adjuvant* slowly leaks the immunogen for exposure to the immune system in small amounts over a period of time.

period. This is the principle of Freund's incomplete adjuvant. However, the immune system is particularly geared to combating microbial infection and its response is stimulated by the presence of components of the bacterial cell wall. Freund's complete adjuvant thus contains bacterial cell wall components, in addition to an emulsifying oil.

The first injection of an immunogen gives a relatively small response and the antibodies produced are of the high molecular weight IgM type (IgM antibodies are comprised of five subunits, each equivalent to a single IgG molecule, joined together). This is known as the primary response. Further inoculations with the same immunogen gives a much greater secondary response, in which mainly IgG-type antibodies are produced (*Figure 98*).

Antibodies are made by a class of white blood cells (leukocytes) known as *B-cells*. In any one animal there are a large number of different B-cells, each capable of making a single type of antibody molecule. Each B-cell carries an "advertisement" γ-globulin on its surface and interaction of this molecule with an antigen molecule causes that B-cell to divide into a clone of similar B-cells, a process known as *clonal expansion*. Some of these B-cells mature into antibody-producing *plasma cells* and some remain as memory cells. The existence of an expanded clone of memory cells accounts for the faster and more extensive secondary response.

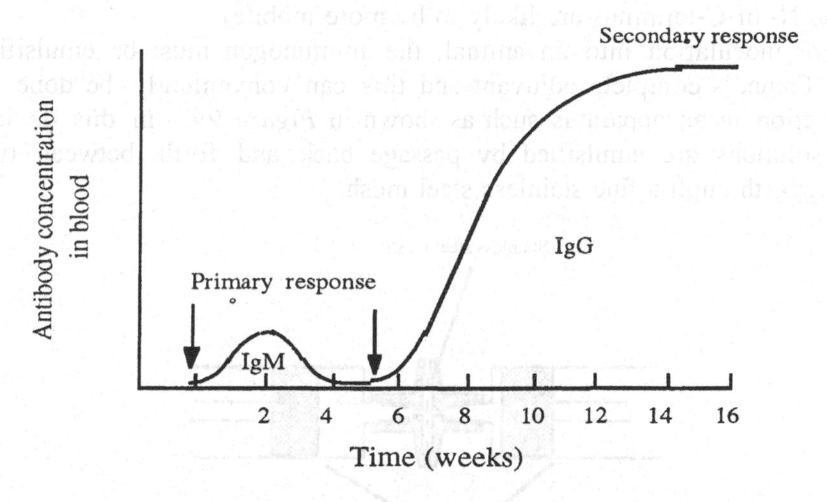


*Figure 98.* Time-course of the immune response.
Arrows indicate inoculations with immunogen.

### 7.2.1 Making an antiserum

An antiserum, or an antibody isolated from it, constitutes a useful reagent in protein biochemistry. However, it is a reagent which is specific to the immunogen/antigen couple and often it must be prepared in-house, especially when a novel protein is under investigation. Preparation of an antiserum starts with the immunogen, which is usually a protein isolated by one or more of the methods described in the previous chapters. The more pure the immunogen, the more specific will be the antibodies which it elicits. For most purposes, therefore, it is best to use as pure an immunogen as possible.

Alternatively, for the production of so-called peptide antibodies, the immunogen might be a synthetic peptide of ten or more residues. The peptide is chosen from the amino acid sequence of the Ag of interest, i.e. the complete protein which it is hoped the Abs will recognise. For peptide antibodies to recognise the whole protein, it is necessary that the peptide sequence chosen be accessible, i.e. it must be on the surface of the protein. This can be readily determined if the 3-D structure of the protein is known. If the 3-D structure is not known, then the peptide can be chosen by analysis of the amino acid sequence of the Ag of interest for hydrophobicity[2,3] (since hydrophilic residues will tend to be on the surface) or mobility (since residues on the surface, and especially at the N- or C-terminus are likely to be more mobile).

For inoculation into an animal, the immunogen must be emulsified with Freund's complete adjuvant and this can conveniently be done by trituration in an apparatus such as shown in *Figure 99*. In this device, the solutions are emulsified by passage back and forth between two syringes, through a fine stainless steel mesh.

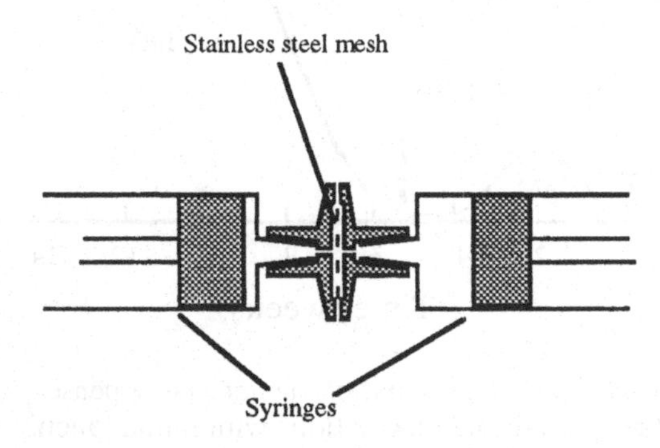


*Figure 99.* A device for emulsifying antibody solution with adjuvant oil.

The triturated immunogen/adjuvant emulsion may be injected subcutaneously in a rabbit, or into the breast muscle of a laying hen. Animals have an idiosyncratic response to immunogens and so it is best to use at least two animals, in case the one is a poor responder - this is provided sufficient immunogen is available, of course. A typical inoculation schedule would involve injection of 50→100 μg of protein per time, first in Freund's complete adjuvant, followed at one, two, four and six weeks thereafter by further inoculations in Freund's incomplete adjuvant. If necessary, further booster inoculations can be given at monthly intervals. In the case of rabbits, blood samples are taken immediately before each inoculation, so that the increase in antibody titre with time can be followed. An illustrative example of the increase in antibody titre with time is shown in *Figure 100*.

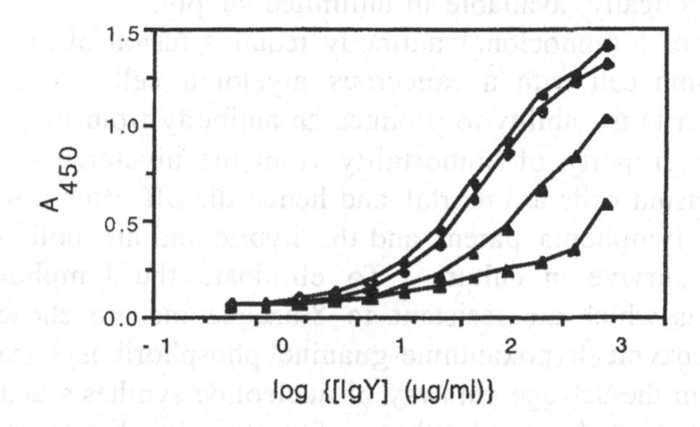


*Figure 100.* A typical ELISA of the progress of an immunisation. The open triangles represent preimmune serum and the other three symbols represent serum at 2, 4 and 6 weeks.

An advantage of using hens for antibody production is that it is not necessary to bleed them in order to harvest the antibodies. It is, of course, necessary to bleed rabbits and this may be done by warming one of their ears with a hot, wet towel, in order to dilate the blood vessels, and nicking the peripheral ear vein with a sharp scalpel blade. About 25 ml can be collected from a rabbit at one time. The blood is best collected into a clean, dry 25 ml conical flask as this has an almost optimal ratio of volume to surface area and the exposed area is also minimal. With a high volume to surface area ratio, the clot can more easily contract away from the vessel walls, thereby obviating tearing of the clot and lysis of the red blood cells. Optimal clot formation is promoted by incubation overnight at 4°C. Ideally, no haemolysis should occur and the serum should be a pale straw colour. It may be harvested by careful aspiration with a Pasteur pipette.

An IgG preparation may be isolated from the antiserum, or IgY isolated from egg yolks, by precipitation with polyethylene glycol[4,5] (See Section 4.5 and Section 8.9).

### 7.2.2 Monoclonal antibodies

A problem with polyclonal antibodies is that every animal has its own unique portfolio of B-cells and so the polyclonal antibodies which an animal produces will be unique to that animal, and will only be available for the lifetime of that animal. Monoclonal antibodies, introduced by Kohler and Milstein in 1975[6], are produced by immortal cell lines and are therefore, theoretically, available in unlimited supply.

Production of a monoclonal antibody requires fusion of an antibody-producing plasma cell with a cancerous myeloma cell. The resulting hybridoma inherits the ability to produce an antibody from its plasma cell parent and the property of immortality from the myeloma cell parent. The parent plasma cells are mortal and hence die off after a number of passages. The lymphoma parent and the hybridoma are both immortal and will both survive in culture. To eliminate the lymphoma cells, lymphoma cells which are resistant to 8-azaguanine are chosen. Such cells lack the enzyme hypoxanthine-guanine phosphoribosyl transferase, a key enzyme in the salvage pathway of nucleotide synthesis and are thus wholly dependent on *de novo* synthesis of nucleotide. Subsequent culture of the lymphoma/hybridoma mixture in aminopterin blocks the *de novo* pathway of nucleotide synthesis and thus eliminates the myeloma cells. The hybridoma cells inherit a functional salvage pathway of nucleotide synthesis from their plasma cell parent, and thus survive in the presence of aminopterin.

Limit dilution of the hybridoma cells enables individual hybridoma clones to be selected, each of which will produce its own monoclonal antibody. As reagents, monoclonal antibodies are essentially not different to other antibodies, except that each monoclonal antibody targets a single epitope. They are also more expensive to produce than polyclonal antibodies.

## 7.3 Immunoprecipitation

The paratope of an antibody and the complementary epitope on an antigen have a very specific stereo relationship with one another. If an antigen contains at least two epitopes, it may form a precipitate upon reaction with its specific (polyclonal) antibodies at optimal proportions. Monoclonal or peptide antibodies, which target only a single epitope, will not form an immunoprecipitate, as they will not be able to form the extended network necessary. The mechanism of formation of an immunoprecipitate is shown in *Figure 101*. Formation of the matrix required for immunoprecipitation requires at least two epitopes, each targeted by a different antibody.

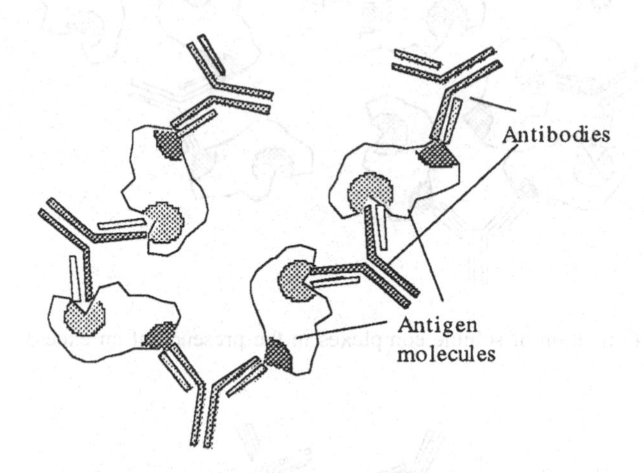


*Figure 101.* The formation of an immunoprecipitate.

Formation of an immunoprecipitate requires optimal proportions of antibodies and antigen and if either the antibody or the antigen is present in excess, then an immunoprecipitate will not form (*Figure 102* and *Figure 103*).

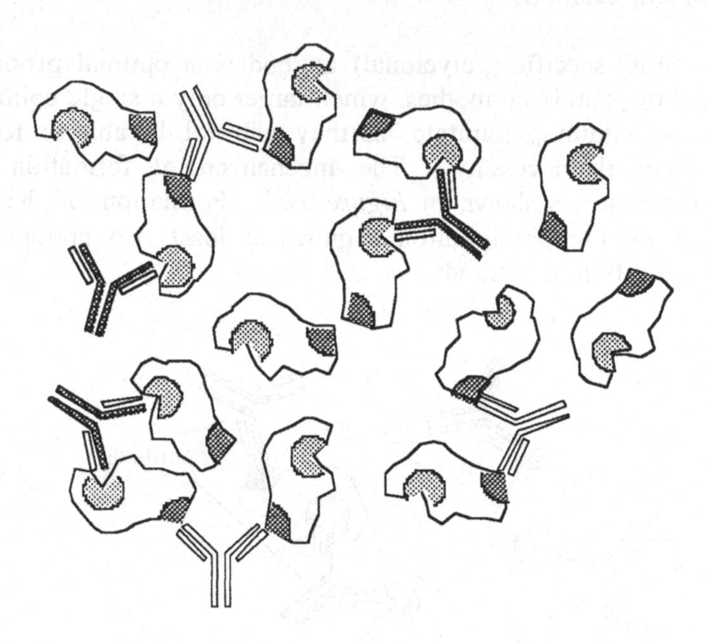

*Figure 102.* Formation of soluble complexes in the presence of an excess of antigen.

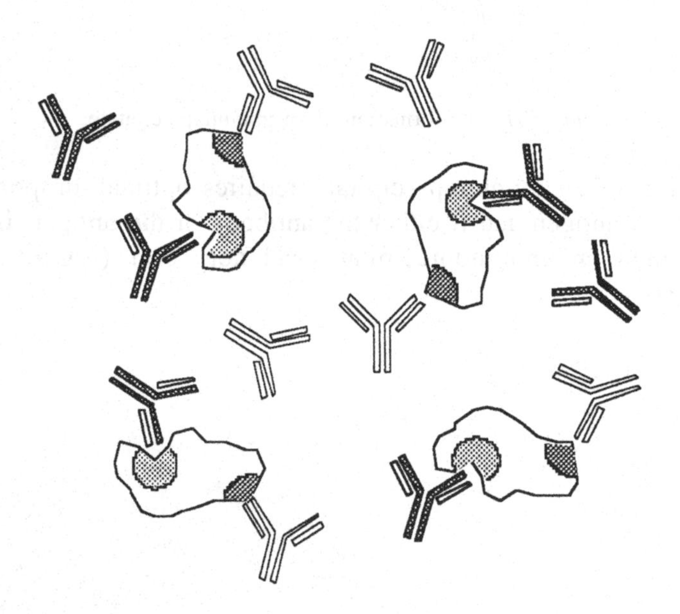

*Figure 103.* Formation of soluble complexes in the presence of an excess of antibody.

The dependence of immunoprecipitation on the proportions of Ab and Ag can be expressed graphically, as in *Figure 104*.

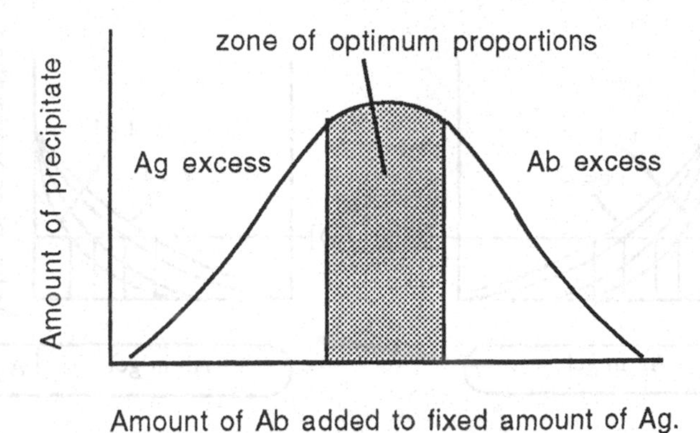


*Figure 104.* Immunoprecipitation at different proportions of Ab and Ag.

Often, with a novel combination of Ag and Ab, the optimal concentrations are not known. This has led to the use of diffusion techniques, where the diffusion of either the Ab or the Ag generates a concentration gradient of that molecule. The optimal concentration will occur somewhere along the concentration gradient and will lead to formation of an immunoprecipitate at that point.

Immunoprecipitation is also affected by pH and in general precipitation will not occur substantially outside of the range pH 5→9.

### 7.3.1 Immuno single diffusion

Immunodiffusion is usually conducted in macroreticular agarose gels, which serve to stabilise the system against buoyancy-driven flow-induced disturbances, including convection, but do not impede diffusion. In immuno single diffusion, the one component is present throughout the gel at a constant concentration, while the other component diffuses into the gel from solution (*Figure 105*).

In immuno single diffusion, the precipitate band moves further into the gel with time, due to the continual diffusion into the gel of the component originally present in solution. The precipitate band also tends to be indistinct as it is spread over an area, with the precipitate band being formed on its leading edge and redissolving on the trailing edge. This "fuzziness" of the bands makes it difficult to determine how many precipitate bands there may be.

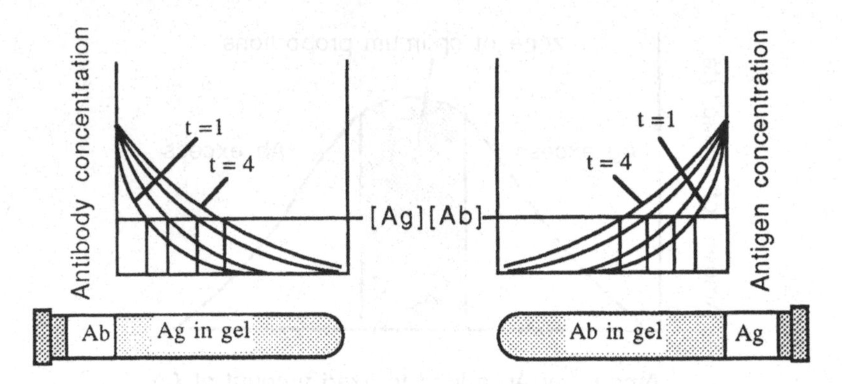


*Figure 105.* Immuno single diffusion.

The Ag or the Ab is present throughout the gel at a constant concentration and the other component is allowed to diffuse into the gel from solution, where it is present at a higher concentration. Immunoprecipitation will occur at the positions where the Ab and Ag are present in equivalent concentrations (indicated by the vertical lines in *Figure 105*). As the one component will continue to diffuse into the gel over time, the position of the immunoprecipitate will move further into the gel with time and the precipitate will not form a sharp line, as it will be re-dissolving on one edge and precipitating on the other.

### 7.3.2 Immuno double diffusion

The limitations of immuno single diffusion can be overcome by using immuno double diffusion. In this case the Ab and the Ag diffuse into opposite ends of the gel and form a precipitate where they meet in equivalent concentrations. With time, the position of the immunoprecipitate will not change, but simply more precipitate will form at the same position as the Ab and Ag continue to diffuse into the gel (*Figure 106*). This is assuming that the Ab and Ag continue to diffuse at the same relative rate, which they will do if the temperature is kept constant.

If there is more than one antigen/antibody couple present, then each of these systems will form their own precipitate band. If the antigens are of different sizes or shapes, they will diffuse at different rates and will form precipitate bands, at different places. (IgG antibodies are all of the same size and gross shape and so will all diffuse at the same rate). Because the precipitate band(s) formed by double diffusion are very sharp, different bands are easily distinguished from one another and so it is relatively easy to determine how many bands have been formed.

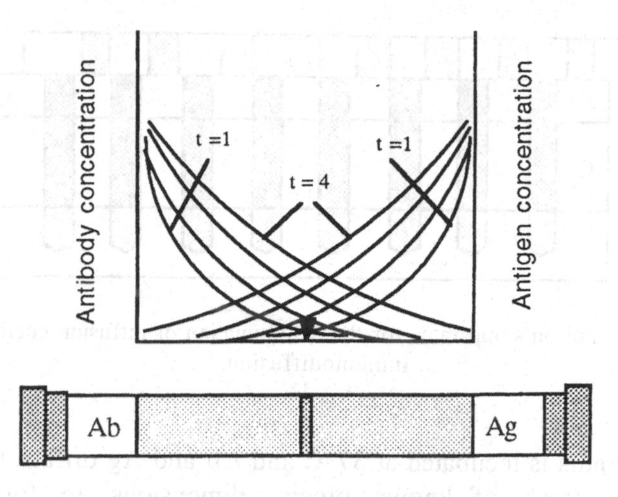


*Figure 106.* Immuno double diffusion.

### 7.3.2.1 Determination of diffusion coefficients

As mentioned above, the development of different precipitin bands in double diffusion is due to differences in the rate of diffusion of different antigens, which is reflected in their different diffusion coefficients. Diffusion coefficients give information about the size and shape of molecules. Because the position of the precipitin band is a function of the diffusion coefficient of the antigen, immunodiffusion can be used to measure diffusion coefficients.

One such method uses a device devised by Polson[7] (*Figure 107*). This consists of four perspex™ blocks, marked A, B, C and D, with holes drilled through three of them and part way into D. The blocks are able to slide relative to one another on joints greased with petroleum jelly.

With the blocks aligned, a serial dilution of Ab can be introduced into the wells in block D after which block D is moved to one side to seal the wells. Molten 1% agarose is introduced into the wells in block C and sealed off by moving blocks A and B to one side, before the agarose sets. Finally, Ag is added to the wells in block B and sealed off by moving block A to one side.

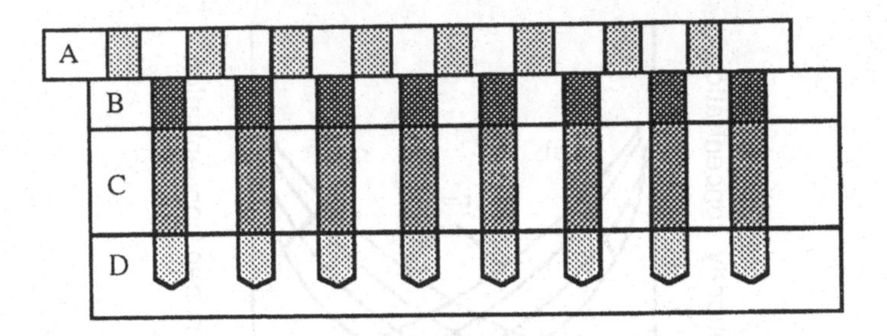


*Figure 107.* Polson's apparatus for the determination of diffusion coefficients by immunodiffusion.

The apparatus is incubated at 37°C and Ab and Ag diffuse through the column of agarose, of known precise dimensions, to form a sharp precipitin band where they meet in optimal proportions. Where the proportions are not optimal the bands are relatively fuzzy (*Figure 108*a).

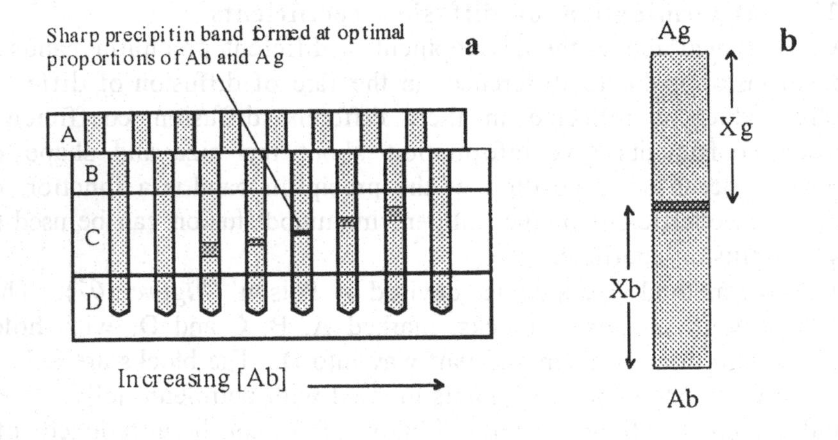


*Figure 108.* Quantitative immunodiffusion in Polson's apparatus.

From the column in which the precipitin band is most sharp, the measurements *Xg* and *Xb* are taken (*Figure 108*b). From these the diffusion coefficient of the Ag can be calculated by substitution in the equation[6]:-

$$\frac{Xg^2}{Xb^2} = \frac{Dg}{Db}$$

Where, $Dg$ = diffusion coefficient of the antigen,
$Db$ = diffusion coefficient of the antibody

Since all IgG molecules have essentially the same gross structure, they will all have the same diffusion coefficient. *Db* is thus a constant (4.6 x $10^{-7}$ $cm^2$ $sec^{-1}$).

Diffusion coefficients can also be measured by molecular exclusion chromatography, since the separating mechanism in this technique is also diffusion dependent. A standard curve of $1/D$ *vs* $K_{av}$, constructed using proteins of known diffusion coefficient *D*, can be used to determine the diffusion coefficients of unknowns[8]. It is useful to have two independent measures of the diffusion coefficient - in this case by immunodiffusion and MEC - as the results from the different methods serve as a check on one another.

## 7.4 Immunoprecipitation methods of historical interest

Immunoprecipitation is the basis of a number of analytical techniques, which although very ingenious, are now mostly of historical interest only. The reason is that immunoprecipitation requires relatively large amounts of antibody and antigen in order to form a visible immunoprecipitate, i.e. it is not a very sensitive technique, and so it has largely been replaced by techniques which involve an amplification step to make the Ab/Ag reaction more easily detected. These will be discussed in Section 7.5.

### 7.4.1 Mancini radial diffusion

A practical, quantitative, single diffusion system is Mancini radial diffusion[9]. In this method the Ab is added to the gel and the Ag to a well cut into the gel. The Ag diffuses into the gel and forms a precipitate where it meets the Ab in optimal proportions, forming a circular precipitin line surrounding the central well. With time, the circular precipitate will grow in diameter until the supply of Ag is exhausted, at which point the growth in diameter ceases. The method gives a quantitative measure of the amount of Ag, because the more there is, the larger will be the diameter of the circle of the precipitin band surrounding the central well. A standard curve can be constructed from the diameters obtained with known concentrations of a standard Ag, and this can be used to determine the concentration of an unknown, from the diameter of its precipitin circle.

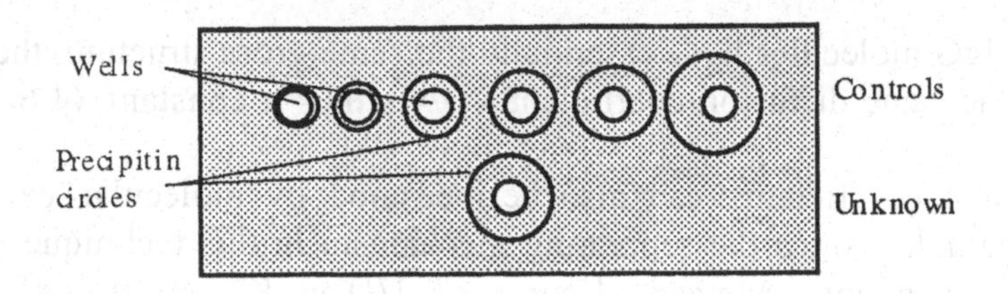


*Figure 109.* Mancini radial diffusion.

### 7.4.2 Ouchterlony double diffusion analysis

Although the concentration gradients formed during immuno double diffusion enable the Ab and Ag to "find" one another at optimal proportions, sometimes the starting concentration of either the Ab or the Ag is out of range, i.e. is too high or too low, and so it is necessary, with an unknown system, to repeat the experiment using different Ab and Ag concentrations. An efficient way of doing this was devised by Ouchterlony[10]. In this method the gel is cast as a layer, 1→1.5 mm thick, on a support (a Petri dish is convenient) and a pattern of wells (*Figure 110*) is cut into it using a die and template. Ab and Ag are added to appropriate wells; typically, Ag may be added to the central well and a serial dilution of Ab added to the surrounding wells.

*Figure 110* shows a single central well surrounded by six circumferential wells, but the Ouchterlony technique is not limited to this arrangement. The pattern can be extended *ad infinitum*, to accommodate different Ab/Ag concentrations.

The serial dilution series is conveniently constructed in the wells of a microtitre plate. A constant volume, say 100 μl, is added to 5 wells of the plate. 100 μl of antiserum is added to the first well and mixed in, 100 μl of this mixture is transferred to the next well, mixed in, and 100 μl transferred to the next well etc., until the last well contains 200 μl of mixture containing 1/32 of the original concentration of Ab.

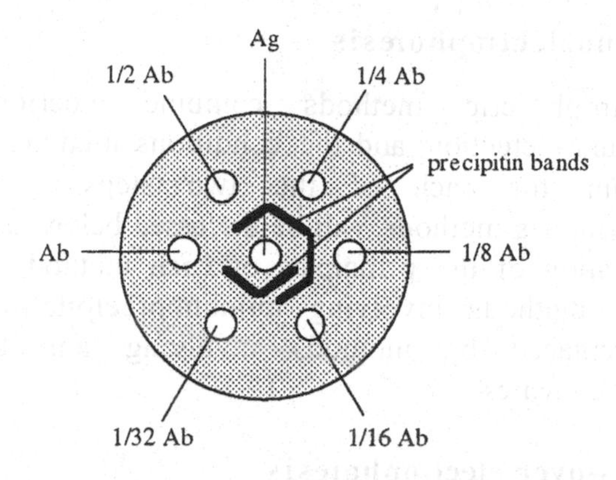


*Figure 110.* Ouchterlony double diffusion.

Three practical points must be borne in mind:-

- Development of the precipitin bands must be done at a constant temperature, usually 37°C, to prevent changes in diffusion rates, which can give rise to indistinct bands.
- To prevent the gel from drying out, it must be incubated in a sealed container with an atmosphere saturated with water vapour. A Tupperware™-type container, containing several sheets of filter paper, saturated with water, is suitable.
- The gel and the Ag and Ab solutions must contain a preservative, such as merthiolate, to prevent microbial growth. A moist environment, at 37°C, in the presence of protein (the Ab and Ag) and carbohydrate (the agarose gel) is ideal for microbial growth.

Ouchterlony double-diffusion can be used in a test of identity of two antigens. If these are identical, their immunoprecipitin lines will fuse into a single line (*Figure 111*a) whereas if they are non identical their precipitin lines will cross (*Figure 111*b). Partial identity is indicated by a fused arc, with a spur (*Figure 111*c).

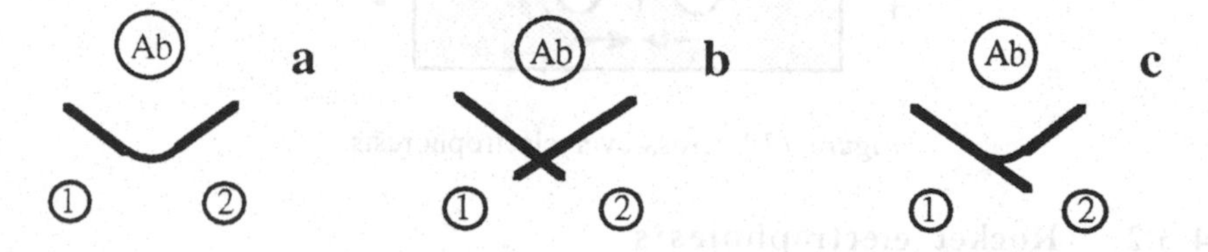


*Figure 111.* Test of identity using immuno double diffusion.

### 7.4.3 Immunoelectrophoresis

Immunoelectrophoretic methods combine electrophoresis with subsequent immuno-detection, and the conditions used are a compromise of the optimum for each of the two steps. A number of immunoelectrophoresis methods will be discussed below, as it is useful to have an appreciation of the principles of each method, but it must be emphasised that methods involving immunoprecipitation have today largely been replaced by methods involving amplification, most commonly with enzymes.

#### 7.4.3.1 Cross-over electrophoresis

One of the limitations of the Ouchterlony double diffusion system is that a large proportion of the Ab and Ag diffuse in non-productive directions and are thus wasted. Only that proportion of the Ab that diffuses towards the Ag, and *vice versa*, will productively form immunoprecipitate.

It will be recalled from the discussion of paper electrophoresis that due to electroendosmosis the γ-globulins - which are the antibodies - migrate towards the cathode, whereas all of the other serum proteins migrate towards the anode (See Section 6.3.1). For the analysis of blood proteins, this provides a way of ensuring that all of the Ag meets all of the Ab, by a process known as cross-over electrophoresis[11]. In this process, the Ab and Ag are placed in two wells in an agarose gel and subjected to electrophoresis as illustrated in *Figure 112*. In this way all of the Ag encounters all of the Ab and wasteful diffusion is obviated. This may be useful when the amount of Ag available is limited, for example in forensics.

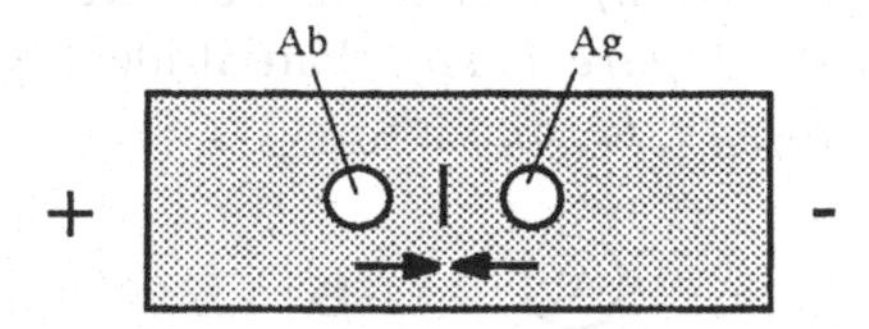


*Figure 112.* Cross-over electrophoresis.

#### 7.4.3.2 Rocket electrophoresis

A limitation of Mancini radial diffusion (Section 7.4.1) was that differences in the diameters of the precipitin rings may be small. The reason for this is that diffusion is in all directions, and a consequence is reduced sensitivity. In the "rocket" electrophoresis method of Laurell[12], diffusion of the Ag is replaced by electrophoresis of the Ag into the gel containing the Ab. Electrophoretic migration of the antigen occurs in one direction only so that immunoprecipitation, instead of occurring in a circle, occurs in a rocket shape (*Figure 113*), hence the name.

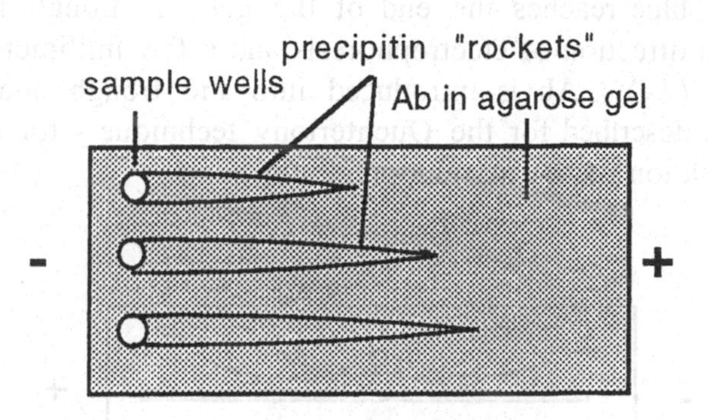


*Figure 113.* Rocket electrophoresis.

An advantage of rocket electrophoresis is that it is more sensitive than Mancini radial diffusion, because, being "pulled out" in one direction only, differences in the length of the rockets are greater and are more easily measured. A standard curve of rocket length *vs* antigen concentration can be constructed and used to determine the concentration of unknowns analysed under the same conditions.

### 7.4.3.3 Grabar-Williams immunoelectrophoresis

Grabar-Williams immunoelectrophoresis[13] is a binary method in which an Ag mixture is first separated by electrophoresis and the separated components are subsequently detected by immuno-diffusion. It may be considered as a development of the Ouchterlony technique, with better resolving power because of the separation of the Ag mixture in the electrophoresis step.

Binary - having two parts

Conditions used have to be a compromise between the requirements of the two stages. For example;-

- Electrophoresis is best performed in a sieving gel, such as polyacrylamide, partly because this restricts diffusion. However, immunodiffusion is *dependent* upon diffusion and so agarose, a non-sieving gel, is used as this does not impede diffusion.
- The requirement for immunoprecipitation restricts the buffer pH, which must not be too far from physiological pH.
- Both electrophoresis and immunoprecipitation require a buffer of low ionic strength.

The sample is introduced into a well cut into an agarose gel, supported on a glass slide, and is separated by electrophoresis. To prevent protein from migrating off the end of the gel, a sample of bromophenol blue may be run in parallel with the protein and the run stopped when the bromophenol blue reaches the end of the gel. A trough is then cut, parallel to the direction of electrophoresis and a few millimeters from the well (*Figure 114*). Ab is introduced into the trough and the gel is incubated - as described for the Ouchterlony technique - for several days to allow formation of the immunoprecipitin bands.

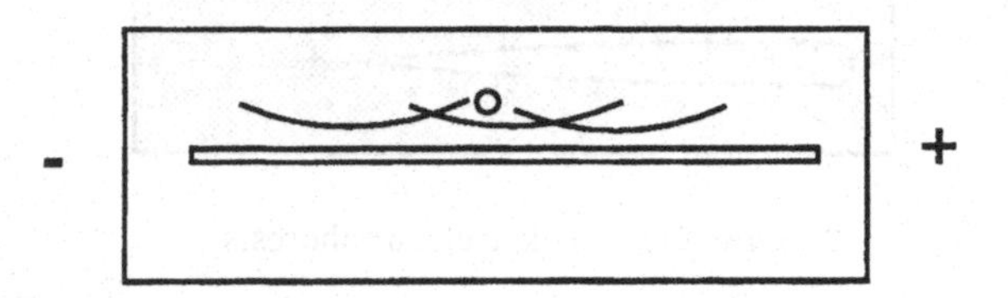


*Figure 114.* Grabar-Williams immunoelectophoresis.

Optimal proportions of Ab and Ag may be established by a prior Ouchterlony analysis, or replicate runs, covering a range of Ab and Ag concentrations, may be carried out.

#### 7.4.3.4 Clarke-Freeman 2-D immunoelectrophoresis

Rocket electrophoresis (Section 7.4.3.2) is a method for the quantitation of Ags by electrophoresis into an Ab-containing gel. In the rocket method different Ags are placed in different wells and either the Ag must be pure or, more usually, a mono-specific antiserum is used. In Grabar-Williams immunoelectrophoresis (Section 7.4.3.3) impure Ag mixtures are first separated by electrophoresis, before being analysed by immuno-precipitation. The Clarke-Freeman technique[14] combines aspects of Grabar-Williams immunoelectrophoresis and of rocket electrophoresis. The sample is first separated by electrophoresis in one dimension, as in the Grabar-Williams technique, and then - at right-angles to the first electrophoresis - is electrophoretically drawn into an Ab-containing gel, as in the rocket technique. Like rocket electrophoresis, the latter stage is a form of cross-over electrophoresis and so an advantage over the Grabar-Williams technique is that all of the Ag meets Ab and there is no wasteful, unproductive, diffusion.

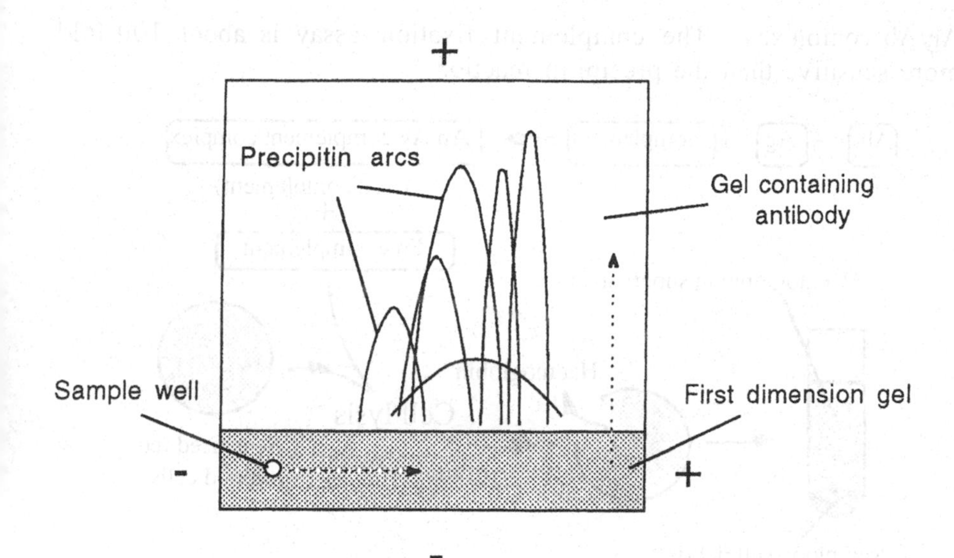


*Figure 115.* Clarke-Freeman immunoelectrophoresis.

Sample in the well is separated by electrophoresis in the first dimension. A gel containing Ab is then cast adjoining the first gel along one of its edges. The separated sample components are then drawn into the Ab-containing gel by electrophoresis in the second dimension.

## 7.5 Amplification methods

At low levels of Ab and Ag, or if monoclonal antibodies are used, no visible precipitation may be formed. In modern methods of immunological analysis, therefore, use is made of amplification methods which enable the Ab/Ag reactions to be visualised and quantitated. Amplification methods will be discussed here in more-or-less historical order, the enzyme and immunogold methods being more recent.

### 7.5.1 Complement fixation

"Complement" is a name given to a group of serum proteins which bind Ag/Ab complexes and cause lysis of cells displaying surface antigens. Complement thus functions *in vivo* as part of the immune response, aimed at the lysis of foreign cells. This activity is exploited in the complement fixation assay, which is a sensitive means of detecting

Ag/Ab complexes. The complement fixation assay is about 100-fold more sensitive than the precipitin reaction.

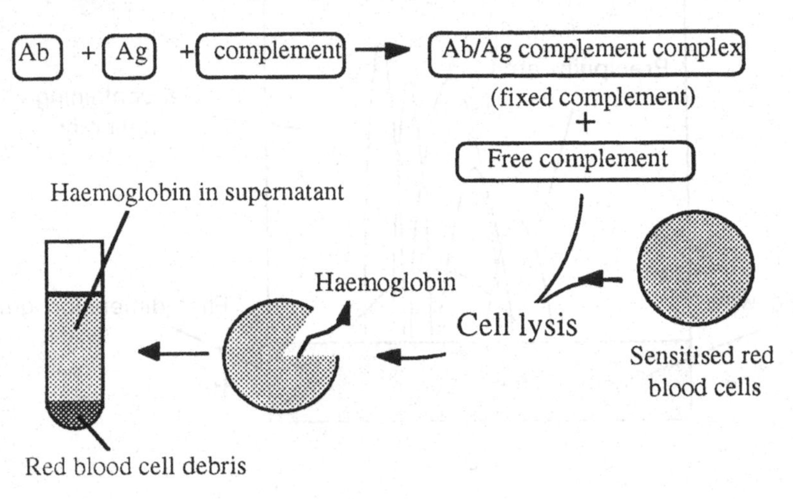


*Figure 116.* The complement fixation assay.

The complement fixation assay depends on the fact that complement is consumed (fixed) by Ag/Ab complexes, making less complement available. The amount of complement left (i.e. not fixed by the Ag/Ab complexes) can be measured by its ability to lyse red blood cells sensitised by antibodies binding to surface antigens. The amount of lysis can be quantitated by measuring the haemoglobin released into the supernatant, by its absorbance at 541 nm.

If all of the complement was bound by Ag/Ab complexes, none would be available to bind to the red blood cells and no lysis, and consequently no release of haemoglobin, would occur. On the other hand, if no Ag/Ab complexes were formed, no complement would be fixed, so all of it would be available to lyse red blood cells and the absorbance of the released haemoglobin would be at a maximum.

In order to measure the amount of a specific Ag present in a complex mixture, it is necessary to first establish a standard curve by measuring the complement fixed when different amounts of Ag are added to a fixed amount of complement and Ab. The standard curve (*Figure 117*) has a shape similar to that of an immunoprecipitation curve (*Figure 104*). Because of the shape of the standard curve, two different Ag concentrations can give the same degree of complement fixation (dotted lines in *Figure 117*). In measuring an unknown, therefore, it is necessary to test several dilutions of the Ag to establish on which side of the curve the Ag concentration falls. The [Ag] can then be read off from the standard curve.

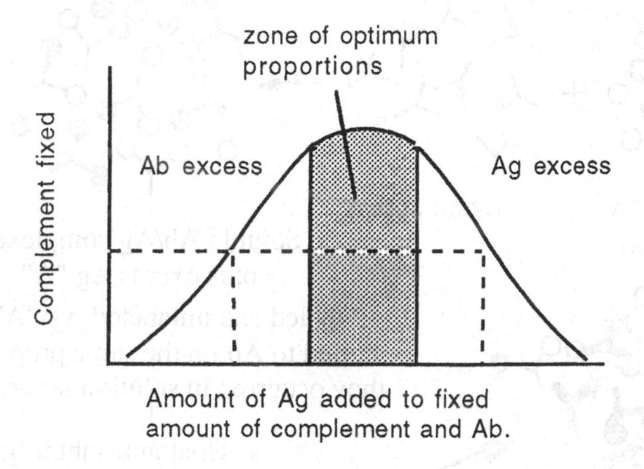


*Figure 117.* Standard curve for complement fixation.

The method has two principal disadvantages:-

- Certain crude mixtures cause haemolysis by mechanisms unrelated to complement, and,
- Some crude mixtures inactivate complement in the absence of the appropriate Ag/Ab complex.

### 7.5.2 Radioimmunoassay (RIA)

Radioimmunoassays (RIAs) combine the sensitivity of radioisotope detection with the selectivity of immunoassays. RIAs can be used to detect molecules that do not fix complement when combined with a specific antibody, for example *haptens* (see p173), and RIAs are mostly used for the assay of small molecules. Examples of compounds that can be assayed by RIA are peptide hormones, steroids (such as testosterone), cyclic AMP, morphine, digitalis etc.

The principle of the RIA is the same as that of the chemical assay technique known as an *isotope dilution assay*. In an isotope dilution assay a known amount of a radioisotope, of known specific activity (i.e. radioactivity per unit mass), is added to a sample and a pure sample of the element of interest is subsequently extracted and its specific activity determined. The decrease in the specific activity of the isolated element is due to the presence of the non-radioactive isotope which dilutes the radioactive isotope. From the extent of this dilution, the concentration of the endogenous, non-radioactive isotope can be determined.

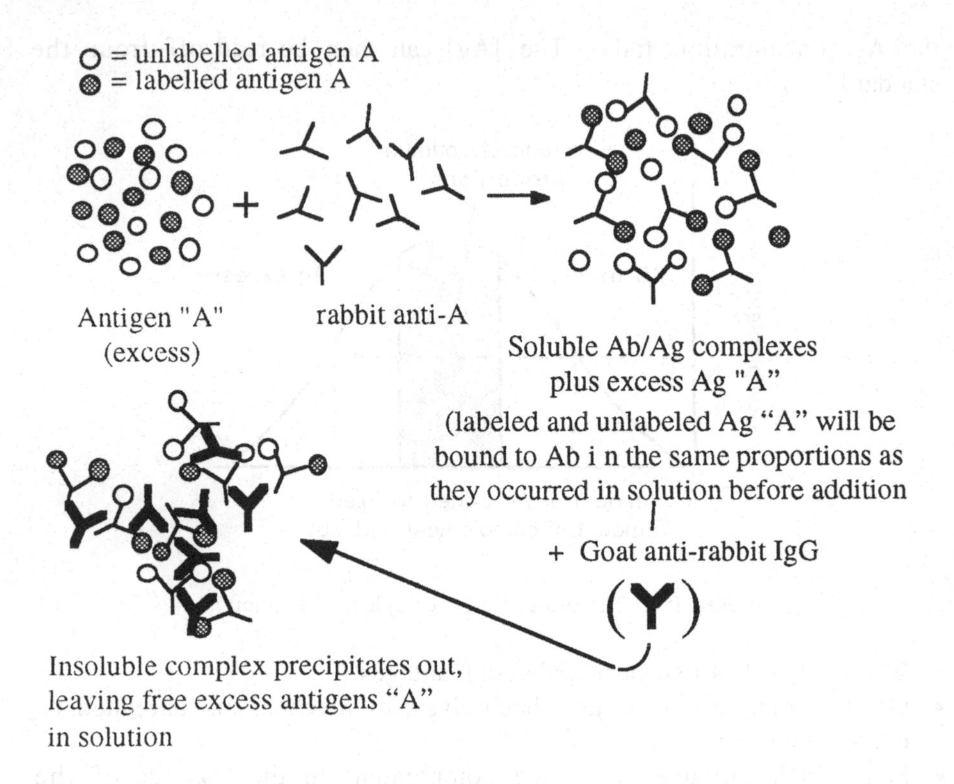
○ = unlabelled antigen A
● = labelled antigen A
Antigen "A"
(excess)
rabbit anti-A
Soluble Ab/Ag complexes
plus excess Ag "A"
(labeled and unlabeled Ag "A" will be
bound to Ab i n the same proportions as
they occurred in solution before addition
+ Goat anti-rabbit IgG
Insoluble complex precipitates out,
leaving free excess antigens "A"
in solution

*Figure 118.* Radioimmunoassay.

A RIA is illustrated in *Figure 118*. In a RIA, a radio-labelled antigen is used to dilute an unknown amount of an unlabelled antigen, present in the sample. A non-saturating amount of, say, rabbit antibody to the antigen is added (i.e. the antigen must be in excess). The labelled and unlabelled antigens will bind to the antibodies in the same ratio as they are present in the sample. The Ab/Ag complexes can then be precipitated, for example by addition of a goat anti-rabbit IgG antibody. The radioactivity in the precipitate will be inversely related to the amount of the antigen originally present in the sample. The concentration of the unknown antigen in the sample can thus be measured by reference to a straight line standard curve in which the % inhibition is plotted against log[non-radioactive Ag].

Radioimmunoassay has the following disadvantages:

- A radioactively labelled Ag may not be available, especially for an antigen which has not been extensively studied.
- Associated with the radioimmunoassay are all of the hazards of radioisotopes, which means that specially equipped laboratories and special licences are necessary.

### 7.5.3 Enzyme amplification

The advantage of enzyme methods over radioisotope methods of amplification is that, because enzymes are safe and biodegradable, no special licences or safety facilities are required, and disposal after use is no problem.

#### 7.5.3.1 Enzyme linked immunosorbent assay (ELISA)

The ELISA method was introduced in its modern form by Engvall and Perlmann[15] and van Weemen and Schuurs[16]. The principle of an ELISA is that an enzyme, linked to an immunoreactive molecule (an antibody or protein A) can be used to detect the presence of an antigen with great sensitivity, due to the amplification achieved by the enzyme catalysed reaction. The method can also be turned about and used to detect Ab. Several different formats of ELISA are possible. Only some concepts pertaining to ELISAs are discussed here. For details on how to conduct an ELISA, a specialist text[17,18] should be consulted.

*The competitive ELISA for measurement of Ag* is conceptually similar to a RIA (Section 7.5.2), in that a labelled Ag competes with an unlabelled Ag for binding to the antibody, but in this case the label is an enzyme (*Figure 119*).

Antibody is immobilised by adsorption onto the walls of a plastic microtitre plate. Excess antibody is washed off and the enzyme-linked Ag is added to one set of wells while the unknown sample, mixed with enzyme-linked Ag is added to another set. The unknown sample Ag thus serves to dilute the amount of labelled Ag which reacts with the immobilised Ab. After incubation, excess antigen is washed off and a solution of the enzyme substrate is added. The enzyme catalysed reaction generates a colour which can be measured. The intensity of this colour is inversely proportional to the amount of Ag originally present. The microtitre plate contains 96 wells in an 8 x 12 array, which permits the exploration of a number of different combinations of coating and Ab/Ag concentrations, permitting optimisation of the reaction[17,18].

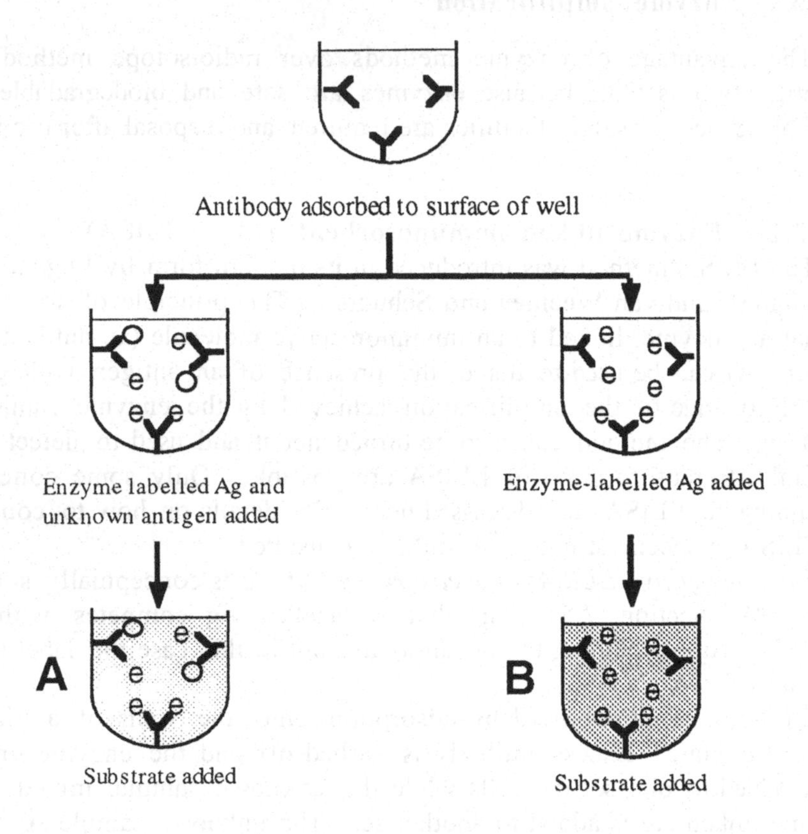


*Figure 119.* A competitive ELISA for measuring [Ag]. Unknown [Ag] is proportional to the absorbance difference between the wells A and B.

In another format, related to immunoblotting (Section 7.5.3.2), the Ag (usually a solution containing a mixture of proteins, including the protein of interest) may be adsorbed onto the walls of a microtitre plate and the Ag of interest subsequently detected with an enzyme-labelled Ab. For increased sensitivity, further amplification may be obtained by using two antibodies, an Ag-specific primary Ab and an enzyme-labeled secondary Ab that is specific for the type of primary Ab. This has the advantage that the secondary enzyme-labeled secondary Ab can be a universal reagent, targeting primary Abs of the same type but with different specificities.

### 7.5.3.2 Immunoblotting

In one ELISA format the principle of amplified detection of an immobilised Ag using an enzyme labeled Ab is used. The Ag, in this case, may be coated onto the walls of a microtitre plate. A similar principle applies to immunoblotting, whereby Ag immobilised on for e.g. a nitrocellulose filter, can be detected using an enzyme-linked antibody system (*Figure 120*). An example is *dot blotting*, in which a small volume of the antigen of interest is dotted onto a nitrocellulose filter[19]. The surrounding protein-binding sites may be blocked with milk proteins, which do not tend to bind proteins non-specifically. The dot can then be probed with a primary antibody, followed by an enzyme-labelled secondary antibody, as in an ELISA. The only difference is that in an ELISA, the enzyme product is soluble, whereas in an immunoblot an insoluble product is necessary. A commonly used combination is horseradish peroxidase as the enzyme label, with 4-chloro-1-naphthol as the substrate. The blue/grey product is insoluble in water[20]. Alternatively, a gold-labelled Ab or protein A may be used[19].

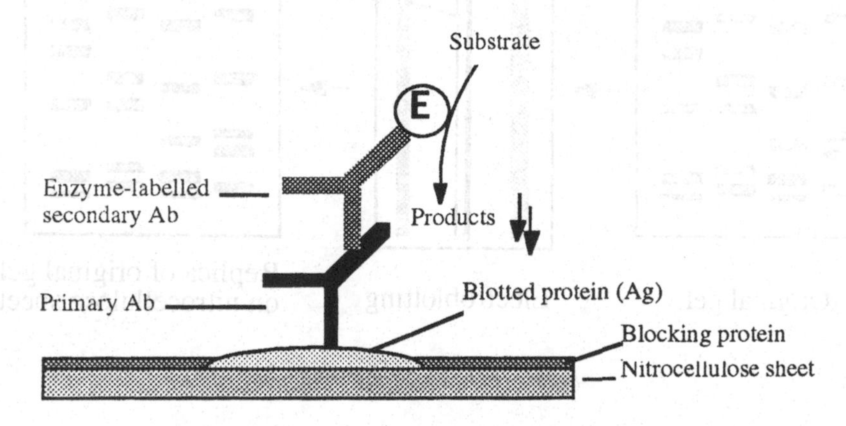


*Figure 120.* A schematic sketch of immunoblotting.

A related technique is *western blotting*[21]. The curious name of this technique is due to a pun. A technique for the blotting of DNA fragments, and probing these with labelled RNA, was named "Southern blotting" after its originator, E. M. Southern. The reverse process, blotting RNA and probing this with DNA fragments, was then named "northern blotting", to indicate that it was an opposite process. The blotting of proteins was subsequently called "western blotting" to indicate a similar process but with different molecules, i.e. in a different direction.

In western blotting, proteins are first separated by SDS-PAGE (p156). The separated proteins are then drawn out of the gel, and blotted onto nitrocellulose, by transverse electrophoresis, forming a replica of the gel separation (*Figure 121*). In this process the nitrocellulose sheet must be on the anodic side of the gel, as protein/SDS complexes are negatively charged and have an anodic migration.

Within the gel the proteins are not accessible to probing by antibodies, but they become accessible after electroblotting onto the surface of a nitrocellulose sheet. Antibodies bound to the blotted proteins can be detected with an enzyme-labelled secondary antibody, as in a dot blot.

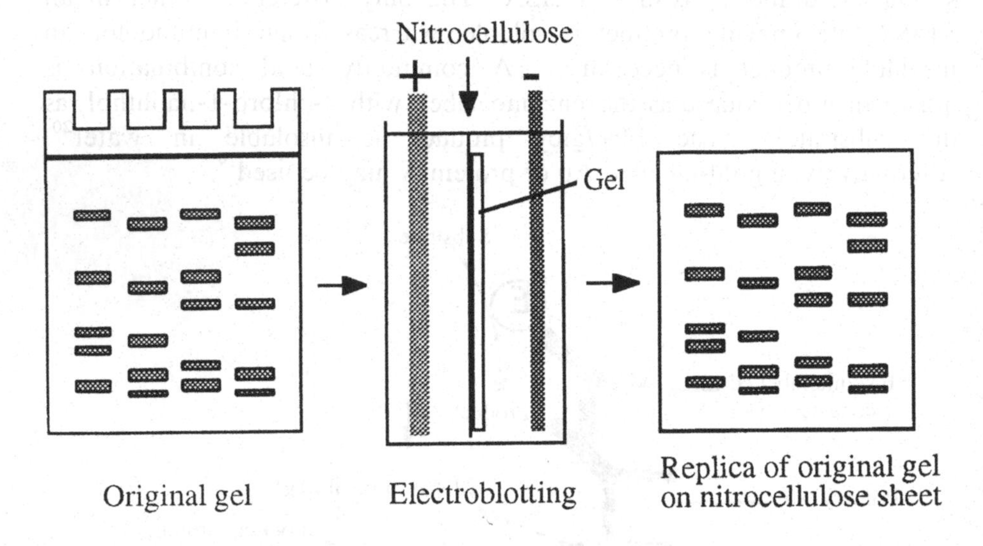


*Figure 121.* Electroblotting.

Treatment with SDS tends to denature proteins, although the effects can be minimised if the sample is not boiled in SDS (see Section 6.8.1). The denatured protein might not be recognised by the antibodies used to probe the blot. In this case a renaturing blot system[22], can be used to advantage. In this technique the transfer buffer does not contain SDS and the SDS may be removed from the gel before the transfer step. The nitrocellulose sheet must be placed on the appropriate side of the gel, depending on the direction of migration of the protein at the pH of the transfer buffer. Normally a high pH is used, giving an anodic migration, but this is not appropriate if the protein is not stable at a high pH.

In a gel of constant composition, proteins separate by virtue of their differential rates of migration, smaller proteins migrating faster than large ones. This same differential will apply to the migration of proteins during the transfer step, i.e. smaller proteins will transfer to the nitrocellulose more rapidly. If insufficient time is allowed for the transfer, the blot will be biased in favour of smaller proteins. This effect can be overcome by running the first electrophoresis in a gradient gel (Section 6.9). In this case the proteins will each reach a point in the gradient where they are about equally impeded by the gel and during the lateral transfer step they will all migrate out of the gel at about the same rate.

Western blotting is useful for determining the presence or absence of a specific Ag in a complex mixture. It is also useful for testing the specificity of an Ab, before this is used in immunocytochemistry, for example. Blotting (not immunoblotting) is also used as a step in the sequencing of a protein band purified by gel electrophoresis. For this purpose the protein is electroblotted onto a polyvinylidene difluoride (PVDF) membrane, the blot excised and transferred into the sequenator.

### 7.5.4 Immunogold labeling with silver amplification

Colloidal gold particles, ranging in diameter from 1 to 30 nm, can be prepared by reducing dissolved gold chlorides with various reducing agents[23]. Proteins bind readily to such particles and stabilise the colloids against a salt challenge. Colloidal gold particles can thus be used as labels, attached to either antibodies or protein A, to form immunogold probes. Such immunogold probes find their greatest use in electron microscopy immunocytochemistry, whereby the subcellular distribution of an Ag of interest may be determined. Colloidal gold particles are very electron dense and show up readily in electron micrographs.

With silver amplification[24], immunogold probes may also be used at the light microscopy level and for staining immunoblots[19]. In the immunogold-with-silver-amplification (IGSS) technique, the colloidal gold label serves as a nucleation centre for the deposition of metallic silver. This yields a black stain which marks the position of the Ag of interest.

### 7.5.5 Colloid agglutination

Proteins can stabilise colloids. The proteins bind to the colloidal particles and similar charge repulsion between the bound proteins keeps the colloidal particles apart, thus preventing flocculation. Latex beads are commonly used as colloidal suspensions for analysis. A natural system, which is virtually colloidal, is blood, in which the red cells are prevented from aggregating by virtue of their similar surface charges.

Antibodies are divalent and are thus able to simultaneously bind to two similar antigens on two different colloidal particles. Such cross-linking of the colloidal particles causes them to flocculate out of suspension, and this provides a very sensitive method for the detection of Abs specific for the colloid-bound Ag. The method is particularly useful for medical diagnosis in the field. Since flocculation can easily be detected by eye, no sophisticated instrumentation is required. The method gives a simple yes-or-no answer, but it can be made semi-quantitative by dilution of the Ab, until the "definite yes" becomes a "maybe". The dilution at which this happens is inversely related to the initial antibody concentration.

An elegant diagnostic method uses agglutination of endogenous red blood cells as the reporter system[25,26]. Monoclonal Abs are raised against glycophorin, a glycoprotein present on the surface of all red blood cells. These Abs are species specific, i.e. they only recognise glycophorin from a particular species. From these monoclonal Abs, $F(ab')_2$ fragments can be made by proteolysis with pepsin. $F(ab')_2$ fragments consist of the two Fab arms of the Ab, bound together by disulfide bridges. The presence of the disulfide bridges is useful as these can be reduced and subsequently used to conjugate a peptide epitope to the free -SH groups of the two separate Fab' fragments. This generates a specific diagnostic reagent (*Figure 122*).

Addition of this reagent to a drop of blood will cause haemagglutination, if Abs targeting the peptide epitope are present. For example, a person infected with the AIDS virus will, in the early stages, have anti-AIDS virus Abs present in their blood. These Abs will target specific epitopes on the AIDS virus proteins. These epitopes can be identified and corresponding peptides can be synthesised. Conjugation of one such peptide to an anti-human glycophorin monoclonal Ab half-$F(ab')_2$ fragment will generate a specific diagnostic reagent. Addition of an appropriate dilution of this reagent to a drop of a patient's blood will give a yes/no indication of the presence of anti-AIDS virus Abs in the patient's blood. A positive answer is given by the agglutination of the red blood cells. Such diagnostic analyses are useful for screening in the field but they should be followed by confirmatory laboratory-based tests, such as ELISA

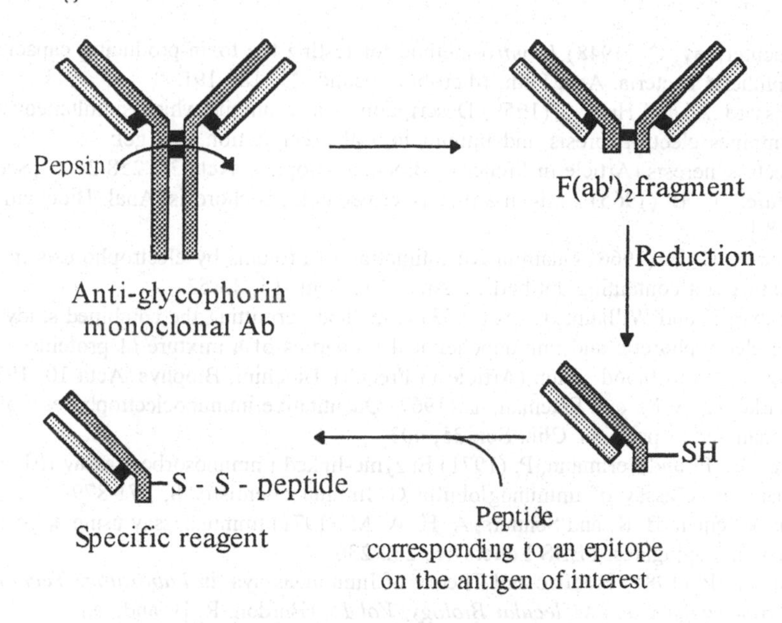


*Figure 122.* Generation of a reagent for detecting the presence of specific antibodies, using endogenous red blood cells as the reporter system.

## *References*


1. Branden, C. and Tooze, J. (1991) in *Introduction to Protein Structure*. Garland Publishing, New York, pp186-191.
2. Hopp, T. P. (1989) Use of hydrophilicity plotting procedures to identify protein antigenic segments and other interaction sites. Methods Enzymol. 178, 571-585.
3. Westhof, E., Altschuh, D., Moras, D., Bloomer, A. C., Mondragon, A., Klug, A., and van Regenmortel, M. H. V. (1984) Correlation between segmental mobility and location of antigenic determinants in proteins. Nature 311, 123-126.
4. Polson, A., Potgieter, G. M., Largier, J. F. Mears, E. G. F. and Joubert, F. J. (1964) The fractionation of protein mixtures by linear polymers of high molecular weight. Biochim. Biophys. Acta 82, 463-475.
5. Polson, A., von Wechmar, M. B. and Van Regenmortel, M. H. V. (1980) Isolation of viral IgY antibodies from yolks of immunized hens. Immunol. Commun. 9, 475-493.
6. Köhler, G. and Milstein, C. (1975) Continuous cultures of fused cells secreting antibody of predefined specificity. Nature 256, 495-497.
7. Polson, A. (1958) A method for determination of diffusion coefficients by the gel-precipitin technique. Biochim. Biophys. Acta 29, 426-428.
8. Andrews, P. (1970) Estimation of molecular size and molecular weights of biological compounds by gel filtration. Methods of Biochemical Analysis 18, 1-53.
9. Mancini, G., Carbonara, A. O. and Heremans, J. F. (1965) Immunochemical quantitation of antigens by single radial immunodiffusion. Immunochemistry 2, 235-254.

10. Ouchterlony, Ö. (1948) *In vitro* method for testing the toxin-producing capacity of diphtheria bacteria. Acta Path. Microbiol. Scand. 25, 186-191.
11. Bussard, A. and Huet, J. (1959) Description of a technique which simultaneously combines electrophoresis and immunological precipitation in a gel: electrosyneresis (Article in French). Biochim. Biophys. Acta 34, 258-260. [See also Laurell, C. B. (1965) Antigen-antibody crossed electrophoresis. Anal. Biochem. 10, 358.]
12. Laurell, C. B. (1966) Quantitative estimation of proteins by electrophoresis in agarose gel containing antibodies. Anal. Biochem. 15, 45-52.
13. Grabar, P. and Williams, C. A. (1953) A method permitting the combined study of the electrophoretic and immunochemical properties of a mixture of proteins: application to blood serum (Article in French). Biochim. Biophys. Acta 10, 193-194.
14. Clarke, H. G. M. and Freeman, T. (1967) Quantitative immunoelectrophoresis of human serum proteins. Clin. Sci. 35, 403.
15. Engvall, E. and Perlmann, P. (1971) Enzyme-linked immunosorbent assay (ELISA) quantitative assay of immunoglobulin G. Immunochemistry 8, 871-879.
16. van Weemen, B. K. and Schuurs, A. H. W. M. (1971) Immunoassay using antigen-enzyme conjugates. FEBS Letters 15, 232-236.
17. Tijssen, P. (1985) Practice and Theory of Immunoassays. in *Laboratory Techniques in Biochemistry and Molecular Biology, Vol 15,* (Burdon, R. H. and van Knippenberg, P. H., eds), Elsevier/North-Holland, Amsterdam.
18. Kemeny, D. M. and Challacombe, S. J. (1988) *ELISA and Other Solid Phase Immunoassays*. John Wiley & Sons, Chichester.
19. Brada, D. and Roth, J. (1984) "Golden blot" - detection of polyclonal and monoclonal antibodies bound to antigens on nitrocellulose by protein A-gold complexes. Anal. Biochem. 142, 79-83.
20. Kerr, M. A. and Thorpe R. (1994) in *Immunochemistry Labfax*. βios Scientific Publishers, Oxford, pp138-141.
21. Towbin, H., Staehelin, T. and Gordon, J. (1979) Electrophoretic transfer of proteins from polyacrylamide gels to nitrocellulose sheets: procedure and applications. Proc. Natl. Acad. Sci. USA 76, 4350-4354.
22. Dunn, S. D. (1986) Effects of the modification of transfer buffer composition and the renaturation of proteins in gels on the recognition of proteins on western blots by monoclonal antibodiies. Anal. Biochem. 157, 144-153.
23. Horisberger, M. (1981) Colloidal gold: a cytochemical marker for light and fluorescent microscopy and for transmission and scanning electron microscopy. Scanning Electron Microsc. II, 9-31.
24. Moeremans, M., Daneels, G., van Dijck, A., Langanger, G. and De Mey, J. (1984) Sensitive visualisation of antigen-antibody reactions in dot and blot immune overlays with immunogold and immunogold/silver staining. J. Immunol. Methods 74, 353-360.
25. Wilson, K. M., Gerometta, M., Rylatt, D. B., Bundesen, P. G., McPhee, D. A., Hillyard, C. J. and Kemp, B. E. (1991) Rapid whole-blood assay for HIV-1 seropositivity using a Fab-peptide conjugate. J. Immunol. Methods 138, 111-119.
26. Wilson, K. M., Catimel, B., Mitchelhill, K. I. and Kemp, B. E. (1994) Simplified conjugation chemistry for coupling peptides to F(ab') fragments: autologous red cell agglutination assay for HIV-1 antibodies. J. Immunol. Methods 175, 267-273.

## 7.6 Chapter 7 study questions

1. Explain the difference between an immunogen and an antigen.
2. Define a "hapten" and describe how anti-hapten antibodies may be raised.
3. An antiserum is described as being a "rabbit (anti-sheep IgG) antiserum". Describe how it would have been prepared.
4. In immunoelectrophoresis,

   The pH must be____________________?
   The ionic strength should be ____________________?
   The temp. must be ____________________?
5. Would it be possible to arrange an immuno-IEF system?
6. In Polson's apparatus for determination of diffusion coefficients by immunodiffusion, an unknown protein formed an optimal precipitin band 1.25 cm from the Ab origin in an apparatus having gel chambers 3 cm long. Calculate the diffusion coefficient of the unknown protein (Db = 4,6 x 10-7 $cm^2$ $sec^{-1}$).
7. Two protein molecules, having the same molecular weight, have different diffusion coefficients. What does this tell one about the proteins?
8. In western blotting, following SDS-PAGE, on which side of the gel should the nitrocellulose membrane be placed, the cathodic side or the anodic side?

# Chapter 8

# Some common practical methods

## 8.1 The Bradford dye-binding assay

Bradford[1], Compton and Jones[2], Read and Northcote[3].

### 8.1.1 Reagents

Dye reagent. Serva blue G (50 mg) is dissolved in 89% phosphoric acid (50 ml). Absolute ethanol (23.5 ml) is added and stirred for 1 h. The solution is made up to 500 ml with dist.$H_2O$, stirred for a further 30 min and filtered through Whatman No. 1 filter paper. The dye reagent can be stored in a brown bottle at room temperature for several months. If precipitation occurs upon storage, the reagent may be filtered and re-calibrated before use.

Standard protein solution. Ovalbumin is dissolved at 1 mg/ml in dist.$H_2O$. This is diluted to 100 μg/ml for the micro-assay.

### 8.1.2 Procedure

Macro-assay. Standard protein solution (0-25 μl), or sample protein, is diluted with dist.$H_2O$ or buffer to a final volume of 100 μl to give the desired concentration levels (0-25 μg). Dye reagent (5 ml) is added to standard solutions, samples and blanks (100 μl), vortexed and allowed to stand for 2 min. The absorbance is read at 595 nm against the buffer blank, for sample, or water for the ovalbumin standard solutions. Assays for a standard curve may be carried out in triplicate at five concentrations of ovalbumin. Plastic cuvettes (3 ml) may be used as the dye-reagent binds to quartz[1], although it can be easily removed with dilute sodium hypochlorite.

Micro-assay. Standard protein solution (0-50 μl of the 100 μg/ml solution, i.e. 1-5 μg) or sample is diluted with dist.$H_2O$ or buffer to a final volume of 50 μl. Dye reagent (950 μl) is added to standard solutions, samples and blanks, vortexed and allowed to stand for 2 min. The absorbance is read at 595 nm against the buffer blank, for sample, or water for the ovalbumin standard solutions, in 1 ml plastic micro-cuvettes as above. A standard curve is generated for each batch of reagent made up, and subjected to linear regression analysis from which protein concentrations may be calculated.

## 8.3 Methods for concentrating protein solutions

### 8.3.1 Dialysis against sucrose or PEG

Where concentration of large volumes of dilute protein solutions is required, dialysis against a substance with a high osmotic pressure in solution, such as sucrose or polyethylene glycol (PEG, $M_r$ 20 000), may be used. Sucrose is inexpensive but is small enough to diffuse through the dialysis tubing. When concentrated sucrose-free protein solutions are required, dialysis against PEG (20 kDa) may be employed.

### 8.3.2 SDS/KCl precipitation

This method is based on the precipitation of SDS-protein complexes by KCl, and is rapid and suitable for the concentration of small volumes, prior to SDS-PAGE. It results in an overall concentration of about 5-fold.

#### 8.3.2.1 Reagents

5% (m/v) SDS. SDS (0.5 g) is dissolved in dist.$H_2O$ (10 ml).

3M KCl. KCl (2.24 g) is dissolved in dist.$H_2O$ (10 ml).

#### 8.3.2.2 Procedure

Sample (100 μl) is diluted with 5% SDS (10 μl) in a 1.5 ml microfuge tube. The solution is mixed by inversion, and 3 M KCl (10 μl) added. The mixture is vortexed and centrifuged (12 000 x *g*, 2 min, RT), the supernatant discarded and the precipitate dissolved in buffer (usually stacking gel buffer).

## 8.4 SDS-PAGE

SDS-PAGE, under reducing and non-reducing conditions is used to examine the purity of protein samples, and molecular weights (of subunits) are usually determined under reducing conditions. This technique is also used with western blotting to analyse samples and to monitor antibody specificity. The inclusion of gelatin in the gels enables SDS-PAGE to be used to monitor the proteolytic activity of samples.

A commonly used method is that of Laemmli[4]. Gels may be routinely stained with Coomassie brilliant blue G-250, or using the rapid method of de Moreno[5], or silver-stained[6] if greater sensitivity is required.

For gels requiring a broader resolving range, the Tris-Tricine gel system[7] may be used. In this system, the mobility of the protein relative to the trailing ion is decreased, and lower molecular weight proteins may be separated.

### 8.4.1 Tris-glycine SDS-PAGE

#### 8.4.1.1 Reagents

Solution A (30% (m/v) acrylamide, 2.7% (m/v) N,N'-methylene-bisacrylamide. Acrylamide (58.4 g) and N,N'-methylene-bisacrylamide (1.6 g) are dissolved and made up to 200 ml with dist.$H_2O$. The solution is filtered and stored in an amber bottle at 4°C.

Solution B 4 x Running gel buffer (1.5 M Tris-HCl, pH 8.8). Tris (36.3 g) is dissolved in about 150 ml dist.$H_2O$, titrated to pH 8.8 with HCl and made up to 200 ml.

Solution C 4 x Stacking gel buffer (500 mM Tris-HCl, pH 6.8). Tris (3.0 g) is dissolved in about 40 ml of deionised dist.$H_2O$, titrated to pH 6.8 with HCl and made up to 50 ml.

Solution D 10% (m/v) Sodium dodecyl sulfate. SDS (50 g) is dissolved in dist.$H_2O$ with gentle stirring and mild heating, cooled and made up to 500 ml.

Solution E 10% (m/v) Ammonium persulfate. Ammonium persulfate (0.5 g) is dissolved in dist.$H_2O$ and made up to 5 ml. Fresh reagent is made up for each experiment.

Solution F Tank buffer [25 mM Tris, 192 mM glycine, 0.1% (m/v) SDS, pH 8.3]. Tris base (12.0 g), glycine (57.6 g) and 10% SDS are dissolved in dist.$H_2O$ and made up to 4 litres. The pH is 8.3.

Crosslinker solution (TEMED). TEMED is used as supplied.

Treatment buffer. [125 mM Tris-HCl, pH 6.8, 4% (m/v) SDS, 20% (v/v) glycerol]. For non-reducing SDS-PAGE, solution C (2.5 ml), solution D (4 ml) and glycerol (2 ml) are made up to 10.0 ml with dist.$H_2O$. For reducing SDS-PAGE, 2-mercaptoethanol (1 ml) is added to the solution prior to it being made up to 10 ml.

Stain stock. [1% (m/v) Coomassie brilliant blue R-250]. Coomassie brilliant blue R-250 (2.0 g) is dissolved in dist.$H_2O$ overnight with stirring. The solution is made up to 200 ml and filtered through Whatman No. 1 filter paper.

Stain. [0.125% (m/v) Coomassie brilliant blue R-250, 50% methanol, 10% acetic acid]. Stain stock (62.5 ml), methanol (250 ml) and acetic acid (50 ml) are made up to 500 ml with dist.$H_2O$.

Destaining solution 1. (50% methanol, 10 % acetic acid). Methanol (500 ml) and glacial acetic acid (100 ml) are made up to 1 litre with dist.$H_2O$.

Destaining solution 2. (5% methanol, 7% acetic acid). Methanol (50 ml) and glacial acetic acid (70 ml) are made up to 1 litre with dist.$H_2O$.

Molecular weight markers. (1 mg/ml standard protein). Bovine serum albumin (1 mg), ovalbumin (1 mg), carbonic anhydrase (1 mg) and cytochrome C (1 mg) are dissolved in treatment buffer (1 ml) and bromophenol blue tracking dye [0.1% (m/v) in stacking gel buffer] (15 µl) is added.

The procedures described below are based on the Hoefer Scientific SE 250 "Mighty Small" and SE 600 apparatus. The amounts of each reagent required to cast two gels (1.5 mm) for each apparatus is shown in Table 7.

*Table* 7. Reagent volumes to cast running and stacking gels for the SE 250 and SE 600 apparatus.

| | SE 250 12.5% T | | SE 250 10% T | | SE 600 10% T | |
|---|---|---|---|---|---|---|
| Reagent | Running gel (ml) | Stacking gel (ml) | Running gel (ml) | Stacking gel (ml) | Running gel (ml) | Stacking gel (ml) |
| A | 6.25 | 0.94 | 5 | 0.94 | 20 | 2.66 |
| B | 3.75 | | 3.75 | | 15 | |
| C | | 1.75 | | 1.75 | | 5 |
| D | 0.15 | 0.07 | 0.15 | 0.07 | 0.60 | 0.2 |
| E | 0.075 | 0.035 | 0.075 | 0.035 | 0.30 | 0.10 |
| TEMED | 0.0075 | 0.018 | 0.0075 | 0.018 | 0.020 | 0.010 |
| dist.$H_2O$ | 4.75 | 4.2 | 6 | 4.2 | 24.1 | 12.2 |
| Volume | 15 | 7 | 15 | 7 | 60 | 20 |

The pore size of the running gel may be varied to achieve optimum resolution between bands of similar molecular weights. Running gel concentrations of 12.5% T are suitable for the molecular weight range 12-75 kDa, while 10% T is suitable for monitoring a slightly lower range.

#### 8.4.1.2 Procedure

Both types of apparatus are assembled according to the manufacturer's instructions. The aluminium and glass plates are washed thoroughly with detergent and water and rinsed with ethanol. The aluminium plates are placed on the gasket of the pod apparatus. Two 1.5Êmm spacers are greased and placed vertically on the outside edges of the plate. On top of this, the glass plates are aligned so that the bottom edges of the aluminium and glass plates, as well as the spacers, are level with the bottom of the pod. The apparatus is secured with the clamps provided. If a gel-casting apparatus is not available, molten agarose solution (0.1%) is poured on a glass plate in two lines which correspond with the bottom of the assembled pod. The pod is placed on the molten agarose in a manner which allows the agarose to be drawn between the glass and aluminium plates by capillary action. Once the agarose has set, it forms a plug which seals the bottom of the pod. The running gel (prepared according to Table 7) is poured into the sandwich, using a large gauge needle, to a level about 3 cm from the top of the glass plate. Care is taken not to trap any air bubbles in the gel during this procedure. The gel is overlaid with dist.$H_2O$ to exclude oxygen, which would prevent the acrylamide from polymerising. Polymerisation normally takes *ca.* 1 h, and is visualised by the formation of a visible gel-water interface. The water is poured off the gel, and the stacking gel buffer is added up to the top of the aluminium plate. The gel is sealed by the insertion of a 15- or

10-well comb, and left to polymerise (20-30 min). Once the polymerisation has taken place, the combs are removed and the wells washed with dist.$H_2O$. The pod is placed in the container and cold tank buffer is poured into the wells and into the cathodic and anodic compartments.

Protein samples are treated with treatment buffer (1:1) and samples to be reduced are boiled for 90 s in a boiling water bath. The treated samples are underlaid into the wells using a fine-tipped Hamilton syringe. The lid is placed on the pod and the pod is connected to a circulating water bath set at about 5°C. The lid is connected to a power supply and the gels are run at a constant 18 mA per gel. When the bromophenol blue tracking dye has reached the bottom of the gel, the power is switched off, the pod disassembled and the gels placed in staining solution for 4 h. Gels are destained with destaining solution 1 overnight, and placed in destaining solution 2 until the background has faded sufficiently to view the bands. Gels are photographed and stored in zipseal bags at room temperature.

The Hoefer SE 600 is assembled according to the manufacturer's instructions. Two glass plates are separated by 1.5 mm spacers and sealed with two long plastic clamps. The plates are placed in the stand and clamped to seal the bottom edge against the rubber of the stand. The running gel mixture, prepared according to Table 7, is poured into the space between the glass plates, to a level about 4 cm from the top. This is overlaid with water and left to polymerise overnight. The stacking gel is added to a level of 2 cm from the top, and an appropriate comb is inserted. The stacking gel takes about 40-50 min to set.

When the gel is set, the cathode compartment is clamped securely onto the glass plates. The apparatus is placed in the tank and tank buffer is poured into the lower chamber, so that no bubbles are present. Sample is applied to the top of the stacking gel. Tank buffer is added carefully to the reservoir to avoid disturbing the sample. The lid is placed on the apparatus and the gels are run at a constant 80 mV. Gels are removed from the pods when the bromophenol blue passes from the bottom of the gel.

### 8.4.2 Tris-tricine SDS-PAGE

#### 8.4.2.1 Reagents

Monomer solution [49.5% (m/v) acrylamide, 3% (m/v) N,N'-methylene-bisacrylamide]. Acrylamide (120 g) and N,N'-methylene-bisacrylamide (3.8 g) are dissolved in dist.$H_2O$ and made up to 250 ml. The reagent is stored at RT in an amber bottle.

Gel Buffer [3 M Tris-HCl, 0.3% (m/v) SDS, pH 8.45]. Tris (90.38 g) is dissolved in 200 ml of dist.$H_2O$, adjusted to pH 8.45 with HCl, and made up to 250 ml.

Anode buffer [0.2 M Tris-HCl, pH 8.9]. Tris (24.22 g) is dissolved in 950 ml of dist.$H_2O$, adjusted to pH 8.9 with HCl and made up to 1 litre.

Cathode buffer [0.1 M Tris-HCl, 0.1 M tricine, 0.1% (m/v) SDS, pH 8.25]. Tris (12.11 g), tricine (17.92 g) and 10% (m/v) SDS (10 ml) are made up to 1 litre with dist.$H_2O$.

Treatment buffer [125 mM Tris-HCl, pH 6.8, 4% (m/v) SDS, 20% (v/v) glycerol]. For non-reducing SDS-PAGE, solution C (2.5 ml), solution D (4 ml) and glycerol (2 ml) are made up to 10.0 ml with dist.$H_2O$. For reducing SDS-PAGE, 2-mercaptoethanol (1 ml) is added to the solution prior to being made up to 10 ml.

#### 8.4.2.2 Procedure

For the Hoefer Scientific SE 250 "Mighty Small" apparatus, the composition of the gels is described in Table 8. Gels are run at 80 V until the dye has entered the separating gel, after which the voltage is increased to 120 V. When the marker dye reaches the end of the gel, the current is stopped, and the gels are removed and stained by either Coomassie or silver staining.

*Table 8.* Preparation of the resolving and stacking gels for Tris-tricine SDS-PAGE

| Reagent | Resolving gel 10% T, 3% C | Stacking gel 4% T, 3% C |
|---|---|---|
| Monomer | 3 ml | 1.5 ml |
| Gel buffer | 5 ml | 0.5 ml |
| Dist $H_2O$ | 14.95 ml | 6.22 ml |
| Ammonium persulfate | 50 µl | 30 µl |
| TEMED | 5 µl | 3 µl |

## 8.5 Serva blue G rapid stain

This following description is of a modification of the Serva blue G silver stain method of de Moreno[5] and is used to visualise bands from preparative SDS-PAGE rapidly, prior to excision from the gel.

### 8.5.1 Reagents

Serva blue G dye reagent [0.25% (m/v) Serva blue G, 50% (v/v) methanol, 12.5% (m/v) TCA. Serva blue G (1.25 g) is dissolved in methanol (250 ml) and dist.$H_2O$ (125 ml) and 50% (m/v) TCA (125 ml) are added with continuous stirring.

40% (v/v) Methanol, 10% (v/v) acetic acid. Methanol (400 ml) and acetic acid (100 ml) are diluted to 1 litre with dist.$H_2O$.

5% (m/v) TCA. TCA (50 g) is dissolved in 1 litre of dist.$H_2O$.

10% (v/v) Ethanol, 5% (v/v) acetic acid. Absolute ethanol (100 ml) is mixed with acetic acid (50 ml) and diluted to 1 litre with dist.$H_2O$.

### 8.5.2 Procedure

Following electrophoresis, strips are excised from the left, right and centre of the preparative gel, and soaked in 40% methanol, 10% acetic acid (100 ml, 15 min), and stained in Serva blue G dye reagent (100 ml, 5 min). The strips are destained in 5% TCA (100 ml, 2 x 10 min) and 40% methanol, 10% acetic acid (100 ml, 10 min). The strips are aligned next to the original gel, areas of the gel corresponding to the stained bands of interest are excised, and macerated for subsequent electro-elution.

## 8.6 Silver staining of electrophoretic gels

For gels where very small amounts of protein need to be visualised, the silver stain technique of Blum *et al.*[6] may be used. This technique boasts greater sensitivity than other silver stain techniques, detecting as little as nanogram amounts of protein. In many silver staining procedures, the pH change causes the non-specific formation of silver salts on the gel, reducing the contrast. By treating the gel with thiosulfate, which complexes and dissolves silver salts, Blum *et al.*[6] found that the background could be reduced.

### 8.6.1 Reagents

Milli-Q deionised water is used for all reagents and rinse steps.

Fixing solution [50% (v/v) methanol, 12% (v/v) acetic acid, 0.2% (v/v) formaldehyde]. Methanol (100 ml), glacial acetic acid (24 ml) and 37% formaldehyde (0.1 ml) are made up to 200 ml with dist.$H_2O$ just before use.

Wash solution [50% (v/v) ethanol]. Absolute ethanol (100 ml) is made up to 200 ml with dist.$H_2O$.

Pre-treatment solution [0.02% (m/v) $Na_2S_2O_3.5H_2O$]. $Na_2S_2O_3.5H_2O$ (0.2 g) is made up to 1 litre with dist.$H_2O$.

Impregnation solution [0.2% (m/v) $AgNO_3$, 0.03% (m/v) formaldehyde]. $AgNO_3$ (0.4 g) and 37% formaldehyde (0.15 ml) are made up to 500 ml with dist.$H_2O$.

Development solution [6% (m/v) $Na_2CO_3$, 0.0004% (m/v) formaldehyde]. $Na_2CO_3$ (12 g), pre-treatment solution (4 ml) and 37% formaldehyde (0.1 ml) are made up to 200 ml with dist.$H_2O$.

Stop solution [50% (v/v) methanol, 12% (v/v) acetic acid]. Methanol (50 ml) and glacial acetic acid (12 ml) are made up to 100 ml with dist.$H_2O$.

### 8.6.2 Procedure

Staining is conducted in clean glass containers, rinsed with chromic acid and ethanol. All steps are carried out on a rocker or orbital shaker. The gel is soaked in fixing solution (1 h), followed by wash solution (3 x 20 min) to neutralise the gel prior to treatment with the acid labile $Na_2S_2O_3$. The gel is treated with pre-treatment solution (1 min), rinsed with dist.$H_2O$ (3 x 20 s) and soaked in impregnation solution (20 min). Gels are again rinsed thoroughly (3 x 20 s), to remove excess $AgNO_3$ from the gel surface and immersed in development solution until bands are evident against a lightly stained background (10 min). The gels may be allowed to develop for longer to visualise any lightly staining bands. Finally, the gel is rinsed in dist.$H_2O$ (2 x 2 min), treated with stop solution (10 min) and stored in sealed plastic bags, in the dark, until scanned or photographed.

## 8.7 Protease zymography

SDS-PAGE can be conducted on polyacrylamide gels in which gelatin has been copolymerised to allow the separation and visualisation of proteolytic enzymes on gels, in positions that correspond to their molecular weights[8].

### 8.7.1 Reagents

1% (m/v) gelatin. Gelatin (0.1 g) is dissolved in 10 ml of dist.$H_2O$, with mild heating.

2.5% (v/v) Triton X-100. Triton X-100 (5 ml) is made up to 200 ml with dist.$H_2O$.

Assay buffer [100 mM sodium acetate, 40 mM cysteine, 1 mM $Na_2EDTA$, 0.02% (m/v) $NaN_3$, pH 5.5]. Glacial acetic acid (2.86 ml) and $Na_2EDTA.2H_2O$ (0.19 g) are dissolved in 450 ml of dist.$H_2O$, adjusted to pH 5.5 with NaOH, $NaN_3$ (0.1 g) is added, the pH checked and readjusted if necessary, and made up to 500 ml. Cysteine-HCl (0.70 g) is added to 100 ml buffer immediately before use.

| $NaN_3$ must not be added to acid solutions as it will release the very toxic gas, $HN_3$ |
|---|

0.1% (m/v) Amido black. Amido black (0.1 g) is dissolved in 100 ml of destaining solution.

Destaining solution [30% (v/v) methanol, 10% (v/v) acetic acid]. Methanol (300 ml) and acetic acid (100 ml) are made up to 1 litre with dist.$H_2O$.

### 8.7.2 Procedure

The procedure for SDS-PAGE gels is modified to include 0.1% (m/v) gelatin in the gels prior to polymerisation. After electrophoresis, the gel is incubated in Triton X-100 (2 x 100 ml, 1 h, RT), followed by incubation in assay buffer, containing 40 mM cysteine (100 ml, 3 h, 37°C). The gel is stained in amido black stain solution (100 ml, 1 h) and destained overnight.

## 8.8 Western blotting

The procedure of western blotting allows proteins separated by SDS-PAGE to be transferred out of the polyacrylamide gel, onto a support matrix where they can be immunologically detected. The method described here is essentially that devised by Towbin *et al.*[8], with a few minor modifications.

### 8.8.1 Reagents

Blotting buffer. Tris base (27.23 g) and glycine (64.8 g) are dissolved in about 3 litres of dist.$H_2O$. To this is added 10% (m/v) SDS (4.5 ml) and CP methanol (900 ml) and the volume made up to 4.5 litres with dist.$H_2O$. The pH is automatically 8.3.

Ponceau S reagent [0.1% Ponceau S in 1% acetic acid]. Glacial acetic acid (1 ml) is diluted to about 90 ml with dist.$H_2O$, Ponceau S (0.1 g) is added, dissolved, and the solution made up to 100 ml. This reagent is made fresh.

Tris buffered saline (TBS) (20 mM Tris-HCl, 200 mM NaCl, pH 7.4). Tris base (2.42 g) and NaCl (11.68 g) are dissolved in about 900 ml of dist.$H_2O$, titrated to pH 7.4 with HCl and made up to 1 litre.

Blocking agent [5% (m/v) fat-free milk powder in TBS]. Fat-free milk powder (5 g) is dissolved in about 90 ml of TBS and made up to 100 ml.

0.5% (m/v) Bovine serum albumin in TBS (BSA-TBS). Bovine serum albumin (0.4 g) is dissolved in about 50 ml of TBS and made up to 80 ml.

0.1% (v/v) Tween 20 in TBS (TBS-Tween). Tween 20 (0.5 ml) is diluted to 500 ml in TBS.

<u>4-Chloro-1-napthol substrate solution [0.06% (m/v) 4-chloro-1-napthol, 0.0015% (v/v) $H_2O_2$]</u>. 4-Chloro-1-napthol (0.03 g) is dissolved in methanol (10 ml), and 2 ml of this solution diluted to 10 ml with TBS. Hydrogen peroxide (35%) (4 µl) is added.

<u>0.1% (m/v) Sodium azide in TBS</u>. $NaN_3$ (0.03 g) is dissolved in TBS (30 ml).

<u>Alkaline phosphatase buffer (100 mM Tris-HCl, 0.5 mM $MgCl_2$, pH 9.5)</u>. Tris (12.1 g) and $MgCl_2.H_2O$ (0.233 g) are dissolved in 800 ml dist.$H_2O$, adjusted to pH 9.5 with HCl and made up to 1 litre with dist.$H_2O$.

<u>Nitro-blue tetrazolium substrate solution [0.003% nitro-blue tetrazolium]</u>. Nitro-blue tetrazolium (0.030 g) is dissolved in 70% (v/v) dimethyl formamide (1 ml).

<u>Bromo-chloro-indolyl phosphate substrate solution (0.0015% bromo-chloro-indolyl phosphate)</u>. Bromo-chloro-indolyl phosphate (0.015 g) is dissolved in dimethyl formamide (1 ml). Immediately prior to use, the nitro-blue tetrazolium solution (1 ml) and bromo-chloro-indolyl phosphate solution (1 ml) are diluted to 100 ml with alkaline phosphatase substrate buffer (100 mM Tris-HCl, 0.5 mM $MgCl_2$, pH 9.5).

### 8.8.2 Procedure

Different types of apparatus may be used. The Hoefer TE Transphor unit, is an example of a liquid blotting system, while the Hoefer Semiphor unit, is an example of a semi-dry blotting system. The latter apparatus is comprised of a platinum-coated niobium anode and a stainless steel cathode, situated between a body cover of inert plastic. The design allows for creation of a uniform electric field between the electrodes and proteins can be transferred with low current and voltage.

Using the Hoefer TE Transphor unit, after separation of proteins by SDS-PAGE (Section 6.8), the polyacrylamide gel is removed, placed on 3 layers of filter paper and the pre-cut nitrocellulose membrane (e.g. Schleicher and Schuell, BA 85, 0.45 µm) is laid squarely on the gel, care being taken to dislodge all air bubbles. Three further layers of wetted filter paper are placed over the nitrocellulose and the complete sandwich is transferred to the TE Transphor electroblotting unit. It is placed vertically between the two electrodes and submerged in blotting buffer, ensuring that the nitrocellulose is on the anodal side. Electroblotting is accomplished in 1.5 h at 200 mA constant current, the apparatus being cooled to 4°C during this time using a circulating water-bath.

Using the Hoefer Semiphor unit, the transfer stack is assembled according to manufacturer's instructions. The mylar mask is placed on the anode. On top of this, 3 pieces of blotting paper (pre-soaked in buffer) are laid, the blotting paper being the same size or smaller than the gel. The pre-wet membrane is arranged on top of the blotting paper, and the gel carefully placed over this. Blotting paper (2-3 sheets) is placed on top of this. At each level of the stack, care is taken to remove air pockets, and components are stacked neatly with all edges parallel. The unit is assembled and a 1 kg weight is placed on the lid to ensure even contact between the electrode and the gel. The current is set to at 0.8 $mA/cm^2$ of gel, with a maximum voltage of 50 V. Transfer is completed in 1 h.

Upon completion of blotting the apparatus is disassembled, the nitrocellulose marked with the positions of relevant wells and hung up to air-dry (1.5-2 h). Alternatively, the membrane is immediately immersed in 0.1% Ponceau S reagent (2 min) and rinsed in dist.$H_2O$ until protein bands are visible. The positions of the molecular weight marker proteins are marked with needle marks and the nitrocellulose is rinsed until the pink colour fades completely. Using scissors the membrane is cut into pieces, corresponding to the sample wells, each piece being placed into a separate container and soaked in blocking agent with gentle rocking (1 h) to saturate additional protein binding sites. The membranes are washed with TBS (3 x 5 min) and antibody diluted in BSA-TBS is added. Controls consist of substitution of a non-immune preparation for an immune antibody for each different antibody, antibody concentration or blotted protein sample tested. After incubation in the primary antibody (2 h or overnight), the nitrocellulose is washed with TBS-Tween (3 x 5 min). The presence of detergent decreases non-specific binding of antibody molecules to the nitrocellulose surface by acting as a further blocking agent, and/or by decreasing non-specific hydrophobic binding. The nitrocellulose is incubated (1 h) in the appropriate secondary antibody-enzyme conjugate diluted in BSA-TBS, again washed in TBS-Tween (3 x 5 min) and incubated in the dark with freshly prepared substrate solution. Incubation in substrate is continued until an optimal colour differential between specifically targeted bands, and non-immune incubations is achieved. The reaction is stopped by briefly rinsing the membranes in 0.1% (m/v) $NaN_3$ in TBS (for HRPO-conjugated secondary antibodies only) and allowing these to dry on a piece of filter paper. Blots are photographed, and stored in the dark to prevent yellowing.

## 8.9 Fractionation of IgG and IgY

A simple and convenient method of purification of IgG and IgY is by precipitation with polyethylene glycol (PEG), a water-soluble linear polymer. Polson *et al.*[9] found that relatively low concentrations of high molecular weight polymers are able to precipitate proteins, but high concentrations of lower molecular weight species are required to effect the same degree of precipitation. The conclusion drawn was that precipitation by PEG is due to molecular crowding. Nevertheless, the concentration of polymer required to precipitate a protein is also a function of the nett charge on the protein.

IgG may be purified from rabbit serum according to Polson *et al.*[9] and IgY from chicken egg yolks according to Polson *et al.*[10] and Rowland *et al.*[11].

### 8.9.1 Reagents

<u>10 mM sodium borate buffer, pH 8.6</u>. Boric acid (2.16 g), NaOH (0.2 g), 37% (v/v) HCl (0.62 ml) and NaCl (2.19 g) are added to dist.$H_2O$ and made up to 1 litre. The pH should automatically be 8.6.

<u>100 mM phosphate buffer, 0.02% (m/v) $NaN_3$, pH 7.6</u>. $NaH_2PO_4$ (13.8 g) and $NaN_3$ (0.2 g) are dissolved in about 800 ml of dist.$H_2O$, titrated to pH 7.6 with NaOH and made up to 1 litre.

### 8.9.2 Isolation of IgG from rabbit serum

Rabbits are bled from the marginal ear vein and the blood allowed to clot overnight at 4°C. Supernatant serum is carefully drawn off the clot, and remaining serum recovered by centrifugation (3 000 x *g*, 10 min, RT) of the clot. The serum is preserved with $NaN_3$, added to 0.02% (m/v). Rabbit serum (1 volume) is diluted with borate buffer (2 volumes). 15% (m/v) 6 kDa polyethylene glycol is dissolved in the protein solution with stirring and the resulting IgG precipitate sedimented (12 000 x *g*, 10 min, RT). The pellet is redissolved in phosphate buffer (3 volumes) and the precipitation procedure repeated to remove remaining contaminants. The final pellet is redissolved in half the initial serum volume with phosphate buffer. In determination of IgG concentration, a 1/40 dilution of IgG in phosphate buffer is made and the absorbance read at 280 nm in a quartz cuvette against a buffer blank. To calculate the protein concentration an extinction coefficient of 1.43 ml/mg/cm is used.

### 8.9.3 Isolation of IgY from chicken egg yolk

Individual yolks are freed of adhering albumin (egg white) by careful washing in a stream of water. The yolk sac is punctured, the yolk volume measured and phosphate buffer, equivalent to 2 volumes of yolk, is added and thoroughly mixed in. Solid PEG (6 kDa) is added to a final concentration of 3.5% (m/v of diluted yolk). The PEG is dissolved with stirring and the mixture centrifuged (4 420 x *g*, 30 min, RT) to separate three phases, a casein-like vitellin fraction, a clear fluid, and a lipid layer on the surface. The supernatant fluid, contaminated with some of the lipid layer, is filtered through a loose plug of cotton wool in the neck of a funnel. The volume of clear filtrate is measured and the PEG concentration increased to 12% (m/v). The precipitated IgY fraction is centrifuged (12 000 x *g*, 10 min, RT), the pellet redissolved in phosphate buffer to the volume after filtration and the IgY again precipitated with 12% (m/v) PEG and centrifuged. The final IgY pellet is dissolved overnight in a volume of phosphate buffer equal to one sixth of the original yolk volume. Immunoglobulin Y concentration is determined using an extinction coefficient of 1.25 ml/mg/cm.

## 8.10 Enzyme-linked immunosorbent assay (ELISA)

One of the simplest and most commonly used ELISAs for the detection of antibodies is a three layer system. Briefly, antigen is coated to the plastic surface of the wells of polystyrene microtitre plates, and the primary antibodies to be quantified allowed to form a complex with the immobilised antigen. After excess antibody has been washed away, the degree or amount of reactivity is quantified with an appropriate detection system. In an ELISA this takes the form of an enzyme conjugated to a secondary antibody which recognises the primary antibody bound to the immobilised antigen. The enzyme reacts with a substrate to yield a coloured product which can be measured spectrophotometrically. This quantitative system complements western blotting, which gives qualitative information about antibody specificity.

An ELISA is commonly used to monitor the progress of immunisation and in cross-species reactivity studies. An ELISA may also be used to determine suitable antibody dilutions for use in western blot analyses.

### 8.10.1 Reagents

Phosphate buffered saline (PBS). NaCl (8.0 g), KCl (0.2 g), $Na_2HPO_4$ (1.15 g) and $KH_2PO_4$ (0.2 g) are dissolved in about 800 ml of dist.$H_2O$, adjusted to pH 7.4 with HCl and made up to 1 litre.

0.5% (m/v) Bovine serum albumin in PBS (BSA-PBS). Bovine serum albumin (0.5 g) is dissolved in PBS and made up to 100 ml.

0.1% (v/v) Tween 20 in PBS (PBS-Tween). Tween 20 (1 ml) is diluted to 1 litre in PBS.

Substrate buffer (150 mM citrate-phosphate, pH 5.0). $Na_2HPO_4$ (2.84 g) and citric acid (2.29 g) are each dissolved in dist.$H_2O$ and made up to 100 ml. The citric acid solution is titrated against the $Na_2HPO_4$ solution (50 ml), to pH 5.0.

Substrate solution [0.05% (m/v) ABTS and 0.0015% (v/v) $H_2O_2$ in citrate-phosphate buffer]. ABTS (7.5 mg) and $H_2O_2$ (7.5 µl) are dissolved in citrate-phosphate buffer, pH 5.0 (15 ml), for one ELISA plate.

0.1% (m/v) Sodium azide in 150 mM citrate-phosphate buffer, pH 5.0. For each ELISA plate $NaN_3$ (15 mg) is dissolved in citrate-phosphate buffer (15 ml).

100 mM Na-borate buffer, pH 7.4. Na-borate (6.18 g) is dissolved in 950 ml dist.$H_2O$, adjusted to pH 7.4 with NaOH and made up to 1 litre.

1 mM Na-acetate buffer, pH 4.4. Glacial acetic acid (1 ml) is diluted to 950 ml with dist.$H_2O$, adjusted to pH 4.4 with NaOH, and made up to 1 litre.

200 mM $Na_2CO_3$ buffer, pH 9.5. Approximately 18.6 ml $NaHCO_3$ (1.68 g/100 ml) is titrated to pH 9.5 with approximately 6.4 ml $Na_2CO_3$ (2.12 g/100 ml).

100 mM sodium periodate solution. Sodium periodate (0.214 g) is dissolved in 10 ml dist.$H_2O$. This is prepared freshly before use.

Sodium borohydride solution (4 mg/ml). Sodium borohydride (8 mg) is dissolved in 2 ml dist.$H_2O$, just before use.

HRPO linked secondary antibodies. The coupling of horseradish peroxidase (HRPO) to immunoglobulin may be effected according to Hudson and Hay[12]. HRPO (4 mg) is dissolved in 1 ml of dist.$H_2O$ and freshly prepared 100 mM sodium periodate (200 µl) is added. The mixture is stirred (20 min, RT), during which it turns a greenish brown, and is dialysed against Na-acetate buffer, pH 4.4, overnight at 4°C. The pH of the solution is raised by the addition of 200 mM $Na_2CO_3$ buffer, pH 9.5 (20 µl) and 1 ml of IgG (8 mg/ml) is added. This is left for 2 h at RT, followed by the addition of fresh sodium borohydride (100 µl) to reduce any free enzyme. This is allowed to stand at 4°C, for 2 h, and dialysed against 100 mM Na-borate buffer overnight at 4°C. An equal volume of 60% glycerol in Na-borate buffer is added and stored at 4°C. The dilution of conjugate to be used is determined by a checkerboard ELISA. The dilution used is that which gives a steep titration curve, above the background, over a serially diluted primary antibody range (usually 1:400 - 1:600 dilution).

### 8.10.2 Procedure

Wells of a microtitre plate (e.g. Nunc Immunoplate) are coated with antigen (150 µl) at predetermined dilutions (as determined by a checkerboard ELISA - e.g. 1.0 µg/ml for many antigens,) in PBS overnight at room temperature. Wells are blocked with BSA-PBS (200 µl) for 1 h at 37°C and washed 3 times with PBS-Tween. Serial two-fold dilutions of the primary rabbit antiserum, IgG or IgY solution in BSA-PBS (starting at 1 mg/ml) are added (100 µl), incubated for 1 h at 37°C and excess antibody washed out 3 times with PBS-Tween. A suitable dilution of sheep anti-rabbit IgG-HRPO conjugate or rabbit anti-chicken IgY-HRPO conjugate in BSA-PBS, is added (120 µl) and incubated (30 min) at 37°C. The ABTS substrate (150 µl) is added and incubated in the dark for optimal colour development (usually 10-20 min). The enzyme reaction is stopped by the addition of 50 µl of 0.1% (m/v) $NaN_3$ in citrate-phosphate buffer and absorbances read at 405 nm in an ELISA plate reader. Titration curves are constructed from the spectrophotometric values.

*References*

1. Bradford, M. M. (1976) A rapid and sensitive method for the quantitation of microgram quantities of protein utilizing the principle of protein dye-binding. Anal. Biochem. 72, 248-254.
2. Compton, S. J. and Jones, C. G. (1985) Mechanism of dye response and interference in the Bradford protein assay. Anal. Biochem. 151, 369-374.
3. Read, S. M. and Northcote, D. H. (1981) Minimization of variation in the response to different proteins of the Coomassie Blue dye-binding assay for protein. Anal. Biochem. 116, 53-64.
4. Laemmli, U. K. (1970) Cleavage of structural proteins during the assembly of the head of bacteriophage T. Nature 277, 680-685.

5. de Moreno, M. R., Smith, J. F. and Smith, R. V. (1985) Silver staining of proteins in polyacrylamide gels: increased sensitivity through a combined Coomassie Blue-silver stain procedure. Anal. Biochem. 151, 446-470
6. Blum, H., Beier, H. and Gross, H. J. (1987) Improved silver staining of plant proteins, RNA and DNA in polyacrylamide gels. Electrophoresis 8, 93-99.
7. Shägger, H. and von Jagow, G. (1987) Tricine-sodium dodecyl sulfate-polyacrylamide gel electrophoresis for the separation of proteins in the range from 1-100 kDa. Anal. Biochem. 166, 368-379.
8. Towbin, H., Staehelin, T. and Gordon, J. (1979) Electrophoretic transfer of proteins from polyacrylamide gels to nitrocellulose sheets: procedure and applications. Proc. Natl. Acad. Sci. USA 76, 4350-4354.
9. Polson, A., Potgieter, G. M., Largier, J. F. Mears, E. G. F. and Joubert, F. J. (1964) The fractionation of protein mixtures by linear polymers of high molecular weight. Biochim. Biophys. Acta 82, 463-475.
10. Polson, A., von Wechmar, M. B. and Van Regenmortel, M. H. V. (1980) Isolation of viral IgY antibodies from yolks of immunized hens. Immunol. Commun. 9, 475-493.
11. Rowland, G. F., Polson, A. and van der Merwe, K. J. (1986) Antibodies from chicken eggs. S. Afr. J. Sci. 82, 339.
12. Hudson, L. and Hay, F. C. (1980) Molecular weights and special properties of immunoglobulins and antigens of immunological interest. in *Practical Immunology*, Blackwell Scientific, Oxford, p347.

# Answers to study questions

## Chapter 1

1. Its weight.
2. Its mass. In a balance, the moments about a fulcrum are equalised. On either side the torque is given by F (=ma) x the moment arm. Since "a" is the same on each side it can be cancelled, leaving the two masses balanced, so an unknown mass can be measured by reference to a known.
3. Yes, the moon experiences a centripetal acceleration towards the earth of 2.72 x 10-3 $m.s^{-2}$, which is what keeps it moving in a circular orbit around the earth.
4. A ship will float when the buoyancy force it experiences is equal to its weight. The buoyancy force and the weight are both forces, which have magnitude and direction. The buoyancy force acts in the opposite direction to the weight, however, and the ship will consequently float when the two forces are of equal magnitude, which will occur when the weight of water displaced is equal to the weight of the ship.
5. 100 metric tons.
6. The Great Lakes are comprised of fresh water, whereas the Mediterranean Sea is comprised of high-salinity sea water. Because fresh water is less dense than sea water, it will take a greater volume of fresh water to equal the weight of the ship and so the ship will float lower in the water in the Great Lakes.
7. If the artificial gravity is doubled, the apparent weight of the ship will double to 200 metric tons. However, the apparent weight of the water displaced will also double and so the ship will float at the same level.
8. The buoyancy force is a consequence of the weight of water displaced. Since in orbit the water (like everything else) is in free fall and is thus essentially "weightless", the immersed sphere would experience no buoyancy force (i.e. the buoyancy force would be zero).
9. Their weight and the drag resulting from their passage through the air.

10. The drag is greatly increased by the parachute (because of its much larger frontal area and its unstreamlined shape) and this results in a much lower terminal velocity.
11. The laden ship will require a greater force to keep it moving at 5 knots, because it will have a greater drag, largely because it has a greater wetted area in contact with the water (and a slightly increased frontal area under the water). Note i) since the ships are identical, we can assume that their shapes are the same, and ii) since the ships are moving at constant speed (i.e. there is no acceleration), the difference in their masses is not relevant.

## Chapter 2

1. Because proteins (especially enzymes) play a key role in biochemical processes and to properly understand these processes we must understand the properties of the proteins carrying them out.
2. Proteins are all very similar, so highly discriminating methods are needed and proteins are labile so only mild methods may be used.
3. Sometimes the question one is endeavouring to address will determine the starting material - otherwise one should select a readily-available material with a high concentration of the protein of interest.
4. 100%
5. No, inevitably there is some loss of material and so real yields are always less than 100%. However, yields are based on activity and if the activity in the homogenate or other early fractions is inhibited, then removal of the inhibitor can give apparant yields of greater than 100%.
6. In general, one should aim for as high a yield as possible, though sometimes the absolute purity is more important than the yield.
7. The specific activity is the units of activity per mg of protein.
8. One should stop when further purification steps achieve no further increase in specific activity or in purity (as measured by analytical methods, such as gel electrophoresis).

## Chapter 3

1. The primary measurement in the assay of an enzyme is the progress curve, i.e. a measure of the amount of product formed per unit time.

2. From the initial slope of the curve, i.e. from the tangent at the origin

3. It never actually stops, but there is no nett reaction after the reaction reaches equilibrium.

4. $V_o$ also doubles.

5. Yes, as described by the Michaelis-Menten equation and the substrate dilution curve.

6. That where $V_o$ is least sensitive to changes in [S], usually >10x Km.

7. Starting from a low temperature, the reaction rate increases with increasing temperature until a temperature is reached where the enzyme starts to denature and the reaction rate decreases. Denaturation is itself a first order reaction with a rate that increases with temperature.

8. To get the proteins into solution where they can be more easily manipulated.

9. By subjecting them to a buffer of low osmotic pressure which will cause water to enter the cells, so that they swell and ultimately burst.

10. Both operate using the principle of laminar flow to induce shearing forces across the diameter of a cell.

11. Due to laminar flow, the wind velocity is zero at ground level and increases with height. A turbine on top of a tower will be exposed to wind of greater velocity.

12. Small particles of dust reside in the boundary layer, which is stationary relative to the surface of the car.

13. A change in velocity with time.

14. i) The one halfway down will reach the bottom first. ii) Yes.

15. A new value for the rpm can be calculated from the expression:-

    specified time x $(\text{specified rpm})^2$ = new time x $(\text{new rpm})^2$

    (the new rpm must be within the limits of the rotor, or you will have to be late).

16. A slow-sedimenting particle, originally near the bottom, will reach the bottom in the same time as a fast-sedimenting particle, originally near the top. By resuspending the pellet in buffer, i.e. 'washing', and re-centrifuging, the number of contaminant, slow-sedimenting, particles in the pellet is reduced. Each 'washing' achieves a further purification.
17. Cream has a density lower than water and, under gravity, will thus experience a positive buoyancy force and will consequently float to the top of the water.

## Chapter 4

1. Freeze-drying removes water. This is useful for i) concentrating the solution and ii) for preservation, since microorganisms require water for their growth and survival.
2. The vapour pressure of water is the pressure of water vapour, in equilibrium with liquid water, measured in a sealed container. The vapour pressure is a function of temperature.
3. It should have short, wide-bore tubing as this minimises the resistance to the flow of water vapour from the sample to the condenser and maximises the rate of freezse-drying - which helps to keep the sample frozen.
4. No, because the vapour pressure approaches zero asymptotically and is virually zero at -50°C. Temperatures lower than -50°C effect an infinitesmal decrease in the vapour pressure of water.
5. A vacuum is any pressure below ambient pressure. An absolute vacuum, which is practically unattainable, is a pressure of zero.
6. With the reduced pressure in the freeze dryer, atomospheric pressure would push the water up to about 7 meters. At that height, the pressure above the column of water could drop to below the vapour pressure of the water at the given temperature, so the water will boil. In doing so, it will reduce the temperature of the water, by removing the latent heat of vapourisation and, eventually, the water might freeze.
7. Dialysis describes the diffusion of small molecules across a semi-permeable membrane. Its rate is affected by the concentration differential across the membrane, the area of the membrane and the distance any molecule has to diffuse in order to reach the membrane.

8. Dialysis against a solution with higher osmotic pressure can be used to effect the concentration of a protein solution. Water is drawn out of the dialysis bag by osmosis.
9. An ultrafiltration membrane rejects large molecules at its surface, whereas molecules larger than the exclusion limit of a dialysis membrane can get some way into the membrane before getting "stuck".
10. Concentration polarisation refers to the differential solute concentration which builds up across a solution undergoing ultrafiltration.
11. Within limits it is essentially independent of the applied pressure, provided this is high enough to effect ultrafiltration in the first place.
12. It is inexpensive and readily available in high purity and the sulfate ion has unique properties that ideally suit its function in salting out.
13. a) Proteins salt out more easily at higher temperatures, b) proteins salt out more readily at pH values below their pI and, c) the effect of [protein] depends on whether the protein is a type I or type II. Type I proteins follow a single precipitation curve and precipitate more readily, the higher their initial concentration. Type II proteins follow different precipitation curves, depending on their initial concentration.
14. Proteins have hydrophilic and hydrophobic patches on their surfaces. In TPP, proteins will equilibrate with the ambient concentrations of solvent (water) and co-solvent (t-butanol), which bind to the hydrophilic and hydrophobic patches, respectively. Upon addition of ammonium sulfate, the sulfate ion sequesters water, leaving a different ratio of solvent to co-solvent. When proteins salt out from a water/t-butanol mixture, they are largely butanolated as there is little water left available to them. In this butanolated state, they are less dense than the water and thus float. In regular salting out, only water is available as a solvent and when salted out the proteins are simply dehydrated and are more dense than the water and thus sink.
15. In TPP, the protein precipitates out of the aqueous phase and into the t-butanol phase and is thus dewatered and desalted. Desalting before, say, IEC is not necessary.

## Chapter 5

1. The distribution coefficient is the ratio of the concentration of solute in the stationary phase divided by the concentration of solute in the mobile phase, at equilibrium.

2. HETP is the height equivalent to a theoretical plate, which is the length of a column in which there is effectively one equilibration of solute between the mobile and stationary phases. The smaller the HETP, the better. The two main factors are i) the size of the stationary phase particles and ii) the mobile phase flow rate.

3. i) The smaller the stationary phase particles, the better. ii) The mobile phase flow rate has an optimum value, which usually lies between 2 and 10 cm h-1, for low pressure liquid chromatography.

4. The smaller the distribution coefficient, the less proportion of time the solute will spend associated with the stationary phase and therefore the faster it will elute. Substance B will thus move more slowly than substance A.

5. To prevent re-mixing of separated solutes after chromatography.

6. It should be rigid, so as to resist compression. It should have a large pore size to permit solute molecules to enter the interior, which gives it a large effective surface area. It should be inert, so that it does not react with buffer components or with sample molecules.

7. DEAE has a positive charge and is thus an anion exchanger which will bind negatively charged proteins. A proteins has a negative charge at pH values above its pI, so DEAE will bind proteins at pH values above their pI values. (This does not preclude a protein binding to a DEAE-group below its pI as binding depends on the protein's local surface charge, whereas the pI is a property of the whole protein).

8. At pH 7.0, both proteins will have a positive charge and a cation exchanger could be used. Both proteins should bind to the exchanger at pH 7 and could be eluted with a gradient of increasing salt concentration. The protein with a pI of 7.6 should have a lesser charge and will be eluted first. Other strategies could also be used.

9. To be a cation exchanger, the substituent group must have a negative charge, i.e. it must be an acid. A strong acidic group is best because it will remain completely ionised over a larger pH range.

10. To ensure that all of the gradient is eluted from the column.

11. An ionic strength gradient is usually better because the ion exchanger substituent groups and the sample proteins can resist changes in pH.

12. a) First the linear flow rate in the first column is calculated.

$$v = \pi r^2 l \text{ cm}^3\text{h}^{-1}$$

Hence,
$$l = \frac{v\, cm^3}{\pi r^2\, cm^2}\ \text{h}^{-1}$$

$$l = \frac{50\, cm^3}{\pi 1.25^2\, cm^2}\ \text{h}^{-1}$$

$$= 10.18 \text{ cm h}^{-1}$$

This is applied to the second column and a new volumetric flow rate is calculated.

$$v = \pi r^2 l \text{ cm}^3\text{ h}^{-1}$$

$$v = \pi\ (0.9)^2 \text{ cm}^2 \text{ x } 10.18 \text{ cm h}^{-1}$$

$$= 25.9 \text{ cm}^3 \text{ h}^{-1}$$

b) First determine what length of column the sample would occupy in the 25 mm column.

$$v = \pi r^2 l$$

$$\text{So, } 30 \text{ cm}^3 \text{ (ml)} = \pi \text{ x } 1.25^2 \text{ cm}^2 \text{ x } l \text{ cm}$$

$$l = 6.11 \text{ cm}$$

Then calculate the volume of this length in the 18 mm column.

$$v = \pi r^2 l$$

$$= \pi \text{ x } 0.9^2 \text{ cm}^2 \text{ x } 6.11 \text{ cm}$$

$$= 15.55 \text{ cm}^3 \text{ (ml)}$$

c) All samples should elute by $V_t$ in molecular exclusion chromatography. However, the trailing edge of a peak eluting at $V_t$ will elute at a volume > $V_t$. Therefore it is safest to elute one-and-a-half column volumes.

$$V = 1.5V_t$$

$$= 1.5 \text{ x } \pi r^2 \, l$$

$$= 1.5 \text{ x } \pi \text{ x } 1.25^2\text{cm}^2 \text{ x } 95 \text{ cm}$$

$$= 700 \text{ cm}^3$$

@ 50 $cm^3$ $h^{-1}$, $= 14$ h

d) 700 ml must be collected in 90 tubes

$$= \frac{700 \; cm^3}{90 \; tubes}$$

$= 7.8 \text{ cm}^3 \text{ tube}^{-1}$ (this is the answer to question e)

@ 50 $cm^3$ $h^{-1}$, $= .16$ h

$= 9.6$ min

15. In a microreticular gel, the fibres are randomly orientated, giving small pores. In a macroreticular gel, the fibres associate to form larger bundles, resulting in greater gel strength and larger pores. Sephadex and polyacrylamide form microreticular gels whereas agarose forms macroreticular gels.

16. By inspection, the unknown has a *MW* between 25 and 45 kDa. A standard curve is drawn, using the data supplied.

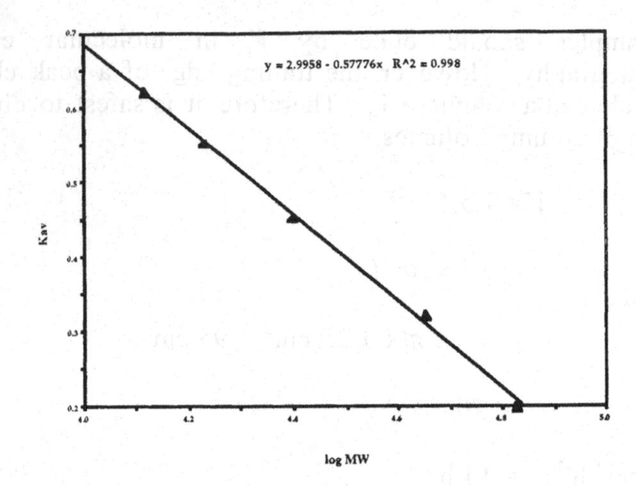


From the regression equation $y = 3.0 - 5.8x$, the *MW* can be calculated as 32 kDa. The subunits must have a *MW* of 16 kDa, which can be calculated to correspond to a $K_{av}$ of 0.562.

Now, $$K_{av} = \frac{V_e - V_o}{V_t - V_o}$$

Therefore, $$V_e = [K_{av}(V_t - V_o)] + V_o$$

Now, $$V_t = \pi r^2 \ell$$

$$= \pi\ (0.75)^2 \text{ x } 50 \text{ ml}$$

$$= 88.4 \text{ ml}$$

And, $$V_o = 0.36\ x\ V_t$$

$\therefore$ $$V_e = .562(88.4-(0.36 \text{ x } 88.4) + (0.36 \text{ x } 88.4)$$

$$= 63.6 \text{ ml}$$

17. 60%

18. A $K_{av} > 1$, indicates that the sample is binding to the gel. In the case of a glucosidase this is probably because the enzyme recognises the glucose units of which the Sephadex is comprised.

19. Hydrophobic interactions are strengthened by high ionic strength. Therefore HI comatography is appropriate after salting out, when the sample contains a high concentration of ammonium sulfate.

20. It is independent of the stationary phase particle size, therefore it is not affected.

21. The phase ratio is the volume of the stationary phase divided by the volume of the mobile phase.

$$K_{av} = \frac{V_e - V_o}{V_t - V_o}$$

$V_s$, the volume of the stationary phase,

$$= V_t - V_o$$

And the volume of the mobile phase $V_o \approx 0.36.V_t$

$$\therefore \text{Phase ratio} = \frac{0.36V_t}{V_t - 0.36V_t} = \frac{0.36V_t}{0.64V_t} = 0.56$$

22.

$$K_{av} = K_d = \text{partition ratio x phase ratio}$$

i.e.

$$0.75 = \text{partition ratio x } 0.56$$

$$\therefore \text{partition ratio} = \frac{0.75}{0.56} = 1.34$$

## Chapter 6

1. Both of these will increase the rate of migration of the protein.
2. As the protein enters the region containing sucrose, the resistance to its migration will increase and its rate of migration will slow down. The protein will also experience a greater buoyancy force, due to the greater density of the sucrose solution.
3. Its electrical resistance will increase. The voltage gradient will get steeper and the protein will migrate more quickly.
4. Smaller ions will have a greater conductivity and this will cause the proteins to have a lower mobility.
5. In most forms of electrophoresis, the support tends to acquire a negative charge, called the zeta potential. Balancing this are $H_3O^+$ ions in the buffer. These cations move towards the negatively-charged cathode, dragging the buffer with them.
6. i) False ii) false iii) false (The anode is in the bottom vessel only in an anionic system. It is in the top vessel in a cationic system).
7. Polyacrylamide gels are more reproducible, but they are not biodegradable, so disposal is a problem, and the monomer is toxic.
8. The pI of the protein.
9. a) The stacking gel prevents density-driven fluid flow which would otherwise occur as the proteins are concentrated into thin discs. b) the running gel improves resolution by gel sieving and by minimising diffusion.
10. Relative mobility vs log molecular weight.
11. Since SDS-PAGE dissociates protein subunits, the measured *MW* will be smaller, unless the protein is not comprised of subunits, in which case the measured *MW* will be the same, so the answer is, smaller-or-equal.
12. Ampholytes are randomly-synthesised, polyamino-polycarboxylic acids, used to generate a pH gradient in IEF and as spacers in isotachophoresis.
13. To improve resolution by establishing very thin starting bands.
14. i) A stable, uniform pH gradient must be established and maintained, ii) the system must be stabilised against disturbances due to buoyancy-driven fluid flows, iii) a system must be devised for application of the

sample, while avoiding pH extremes and, iv) systems must be devised for measurement of the pH gradient and for determination of the positions of the focused bands.

15. a) 30 $V.cm^{-1}$ b) 15 $V.cm^{-1}$ c) it would decrease.
16. First, plot a standard curve of $R_m$ vs log *MW*.

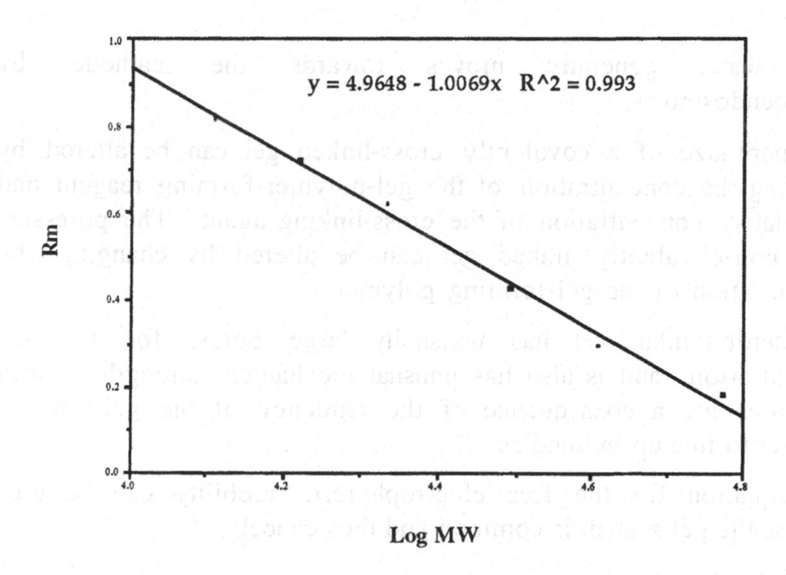


From the regression equation,

$$R_m = 4.965 - (1.007 \times \log MW)$$

calculate the *MW* of the unknown from its $R_m$ of 0.246.

$$0.246 = 4.965 - 1.007 \log MW$$

$$\log MW = (4.965 - 0.246)/1.007$$

$$= 4.686$$

Hence $MW = 48{,}500$

Now calculate $R_m$ values from the reducing SDS-PAGE data. It will be noticed that, under reducing conditions, the standard protein with a $R_m$ of 0.295 (i.e. *MW* 40,500) dissociates into two subunits with Rm values of 0.547 and 0.750, corresponding to *MW*s of about 24,500 and 15,500, respectively. The unknown gives two bands with $R_m$ values of .484 and .578, corresponding to *MW*s of 28,000 and 22,500, respectively. It may be concluded that the unknown protein has a *MW* of about 50,000 and is comprised of two unequal subunits with *MW*s of about 28,000 and 22,000, respectively. (Note the imprecision of the method.)

17. No.

18. Yes, water generally moves towards the cathode by electroendosmosis.

19. The pore-size of a covalently cross-linked gel can be altered by changing the concentration of the gel-polymer-forming reagent and the relative concentration of the cross-linking agent. The pore-size of a non-covalently linked gel can be altered by changing the concentration of the gel-forming polymer.

20. A macroreticular gel has unusually large pores, for the gel concentration, and is also has unusual mechanical strength. Both properties are a consequence of the tendency of the gel-forming polymer to line up in bundles.

21. The equation for the free electrophoretic mobility can be used because the gel system is common and thus cancels.

$$\mu = \text{velocity (voltage gradient)}^{-1}$$

$$= \frac{dx / dt}{dV / dx}$$

$$\text{Therefore,} \quad \frac{35 \text{ mm}}{60 \text{ min}} \times \frac{dx}{90 \text{ V}} = \frac{50 \text{ mm}}{T \text{ min}} \times \frac{dx}{150 \text{ V}}$$

$$\text{Hence,} \quad T = 50 \times \frac{dx}{150} \times \frac{90}{dx} \times \frac{60}{35} \text{ min}$$

$$\approx 52 \text{ min}$$

22. The field lines will be curved outwards, going from the gel into the wider wick. As the field lines are always at right angles to the iso-voltage contours, this implies that the iso-voltage contours will be closer together (i.e. the voltage gradient is steeper) on the inside of the curve. The steeper voltage gradient will cause the proteins on the inside of the curve (i.e. those on the outer edges of the gel) to move more quickly than those where the field lines are straight (i.e. in the middle of the gel). The nett result is that the protein bands become distorted as shown in the figure.

## Chapter 7

1. An immunogen elicits an immune response when injected into an animal. An antigen reacts with an antibody due to the stereo-complementarity of one of its epitopes with the paratope of the antibody.
2. A hapten is a molecule which is too small to act as an immunogen, by itself, but which can nevertheless react with anti-hapten antibodies. These antibodies are elicited by injecting the hapten, conjugated to a protein molecule.
3. IgG would have been isolated from sheep serum and inoculated into a rabbit, with adjuvant, following an inoculation protocol stretching over, say, 8 weeks. The rabbit blood would have been collected, clotted and the serum decanted.
4. a) close to physiological pH, b) low, c) constant.
5. In principle, it should be possible. The only different problem to be overcome would be to eliminate the effect of the pH gradient on the antibody/antgen reactions.
6. The relevant equation (see Section 7.3.2.1) is :-

$$\frac{Xg^2}{Xb^2} = \frac{Dg}{Db}$$

Where, $Dg$ = diffusion coefficient of the antigen,
$Db$ = diffusion coefficient of the antibody
Xg = distance from antigen well to precipitin band
Xb = distance from antibody well to precipitin band

Thus, $Dg = 4.6 \times 10^{-7} \times \frac{1.75^2}{1.25^2}\ cm^2.s^{-1}$

$= 9.02 \times 10^{-7}\ cm^2.s^{-1}$

7. The two molecules have different shapes.
8. SDS-protein complexes have a negative charge and will thus migrate to the positive anode. The nitrocellulose should thus be placed on the anodic side.

## Further sources of information.

**Computer-based exercises in protein isolation:-**

- Amersham Biosciences - www.amershambiosciences.com - search for Protein Purifier software
- www.booth1.demon.co.uk/archive/

**Ultrafiltration:-**

- www.millipore.com

**Magnetic beads:-**

- www.dynalbiotech.com/

**The Parr bomb:-**

- www.parrinst.com

**Homogenisers:-**

- www.virtis.com
- www.ikausa.com (Ultra turrax)
- www.waringproducts.com (Waring blendor)
- www.brinkmann.com (Polytron)

**General protein methods:-**

- www.amershambiosciences.com
- www.bio-rad.com
- www.piercenet.com/proteomics
- www.clontech.com/proteomics

**HPLC:-**

- www.waters.com
- www.lcresources.com/software/software-html
- www.beckmancoulter.com (look at e-labnotebook)
- www.phenomenex.com
- www.sigmaaldrich.com

**Additional reading.**

Deutscher, M.P. (Ed) (1990) *Guide to protein purification.* Methods in Enzymology Vol. 182

Scopes, R. K. (1994) *Protein Purification: Principles and Practice. 3rd Ed,* Springer-Verlag, New York.

Janson, J.-C. and Rydén, L. (Eds) (1998) *Protein purification: principles, high resolution methods, and applications. 2nd Ed.* Wiley, New York.

Wood, E.J. (Ed) (1989) *Practical biochemistry for colleges.* Pergamon Press

Roe, S. (2001) *Protein purification applications - a practical approach.* OUP.

Amersham Biosciences also produce a number of handbooks, including:-

- *Protein purification handbook*
- *Gel filtration: principles and methods*
- *Ion exchange chromatography: principles and methods*

- *Affinity chromatography: principles and methods*
- *Hydrophobic interaction chromatography: principles and methods*
- *Antibody purification handbook*
- *Recombinant protein handbook: amplification and simple protein purification*
- *Expanded bed adsorption: principles and methods*
- *Reverse phase chromatography: principles and methods*
- *Chromatofocusing with PBE*
- *Sephadex LH-20: chromatography in organic solvents*
- *Column packing* (video)

# Index